BRUCE McBRA

AML C

W9-BRS-147

Praise For *The Customer-Driven Company*

"*The Customer-Driven Company* is terrific! It exceeded my expectations because there are so many books written about this subject, it is difficult to anticipate much uniqueness in the next offering. Mr. Whiteley has put together an educational guide that takes the quality improvement process out of the usual cliché-driven theory and brings it into real implementable action steps. If businesspeople will open the cover, they will find great value."

—Ron K. Glover, President, *Dun & Bradstreet Information Services*

"I've been looking for *one book* that combines all the best quality and customer-service ideas together. I've finally found it in *The Customer-Driven Company*. It's become my handbook."

—Stew Leonard, Jr., President, *Stew Leonard's*

"A great customer-service reference book that we can all learn from. As Richard Whiteley so eloquently writes, customer-driven companies will be the only survivors in the '90s."

—Carl Sewell, author of *Customers for Life*

"Quality demands action. Theory alone won't move your firm an inch toward quality improvement. Richard Whiteley provides the reader with helpful guideposts along the way."

—Charles M. Cawley, President and CEO, *MBNA America*

"Richard Whiteley's insights on quality and customer focus are presented in a very readable and useful way. His many specific suggestions on how to become customer-driven make this a valuable addition to a crowded field."

—Richard E. Heckert, former Chairman, *Du Pont*

"Whiteley's roadmap guides anyone serious about product and service quality through a series of practical benchmarks which, in practice, can point to competitive success."

—Sarah M. Nolan, President, Investment and Insurance Services Group, *AMEX Life Assurance Company*

"When it comes to customer-focus techniques, Forum is the best in the world."

—Arthur Taylor, Dean, *Fordham School of Business* , and former President and Director, *CBS, Inc.*

"Forum is the definitive source for thousands of executives who want to know more about the 'whys' and 'hows' of customer focus."

—Richard Atlas, Partner, *Goldman Sachs*

"Right on the money. . . . Companies who deliver both product quality and service quality will be the survivors in the '90s. . . . Mr. Whiteley's 'Action Points' are well worth using."

—Donald Woodley, President, *Compaq Canada, Inc.*

"I really enjoyed reading this book. Dick Whiteley rightfully demonstrates that 70 percent of the reasons why customers switch suppliers or brands has nothing to do with the product or its price. It is the little things, such as care, courtesy, and personalized attention that make the difference. . . . Perhaps the best part of the book is its research-based roadmap for organizations to 'walk the talk' if they want to become customer-driven."

—Jagdish N. Sheth, Robert E. Brooker Professor of Marketing, *University of Southern California*

The Customer-Driven Company

Richard C. Whiteley

The Forum Corporation
America's Leading Experts on
Customer-Focused Quality

THE CUSTOMER-DRIVEN COMPANY

Moving from Talk
to Action

Addison-Wesley Publishing Company, Inc.
Reading, Massachusetts   Menlo Park, California   New York
Don Mills, Ontario   Wokingham, England   Amsterdam   Bonn
Sydney   Singapore   Tokyo   Madrid   San Juan   Paris
Seoul   Milan   Mexico City   Taipei

Library of Congress Cataloging-in-Publication Data

Whiteley, Richard C.
    The customer-driven company: moving from talk to action / Richard
C. Whiteley.
      p.   cm.
    Includes bibliographical references and index.
    ISBN 0-201-57090-4
    1. Customer relations. 2. Customer service. I. Title.
HF5415.5.W56   1991
658.8'12—dc20
                                     90-23765
                                         CIP

Jacket design by Steve Snider
Jacket illustration by James Kaczman
Text design by Joyce C. Weston
Set in 10-point Sabon by ST Associates, Wakefield, MA

1 2 3 4 5 6 7 8 9-MW-9594939291
First printing, February 1991

Addison-Wesley books are available at special discounts for bulk purchases by corporations, institutions, and other organizations. For more information, please contact:

Special Markets Department
Addison-Wesley Publishing Company Inc.
Route 128
Reading, MA 01867
(617) 944-3700 x2431

To Jeffrey, Matthew, and Philip
My Sons, My Friends, My Teachers

■ SPECIAL THANKS

We would like to offer special thanks to the following companies who have contributed the knowledge base from which this book has been created.

ÆTNA Life Insurance & Annuity
   Company
American Express Travel Related
   Services Company, Inc.
AMP Incorporated
Avco Insurance Services
Avon Products Inc.
Bank of Boston Corp.
Bechtel Group, Inc.
Bell Atlantic Mobile Systems, Inc.
Bell Atlantic TriCon Leasing
   Corporation
Bolt Beranek and Newman Inc.
British Railways Board
Budget Rent A Car Corporation
Campbell Sales Company
Charles Schwab & Co., Inc.
Chemical Banking Corporation
Compaq Canada Inc.
Corning
CUNA Mutual Insurance Group
du Pont Company
Eastman Kodak Company
Federal Reserve Bank of Boston
Fidelity Investments Institutional
   Services
General Electric

General Motors
Gillette Company Stationery
   Products Group
Goldman Sachs
Harris Bank
Home Insurance Company
John Hancock Mutual Life
   Insurance Company
Johnson Hill Press
Industrial Indemnity Company
KPMG Peat Marwick
Labatt's Breweries of Canada
Levi Strauss & Co.
Lincoln National Corporation
Matsushita Services Company
Midland Bank PLC
Motorola, Inc.
MSD AGVET, Division of Merck
   & Co., Inc.
New York Life Insurance Co.
N. & P. Building Society
Pepperidge Farm, Inc.
Plymouth Rubber Company, Inc.
Public Service Electric & Gas Co.
R.R. Donnelley & Sons - Book
   Group
Royal Bank of Canada

Sea-Land Service, Inc.
Sharp Electronics Corporation
State Mutual Companies
Tennant
United Technologies Corporation
US West Inc.
Westinghouse Electric Corporation
Westvaco
Xerox Canada Ltd.

Contents

The Missing Ingredients   *1*

1. Create a Customer-Keeping Vision   *21*
2. Saturate Your Company with the Voice of the Customer   *39*
3. Go to School on the Winners   *69*
4. Liberate Your Customer Champions   *87*
5. Smash the Barriers to Customer-Winning Performance   *117*
6. Measure, Measure, Measure   *149*
7. Walk the Talk   *177*

A Final Word   *209*

   Appendix: The Research   *215*

   Customer-Focus Toolkit   *219*

   1. The Characteristics of a Customer-Driven
      Company: A Self-Test   *220*
   2. Tools for Developing a Vision   *226*
   3. Tools for Smashing Barriers to Customer-
      Winning Performance   *229*

Acknowledgments   *295*

Index   *301*

Reader's Questionnaire   *307*

Always deal with complaints before they're made.
  —*Timothy Firnstahl,*
    Chief Executive
    Satisfaction Guaranteed Eateries[1]

The Missing Ingredients

Picture this. True story. You've just spent twenty-five hours flying from Boston to Singapore. You escape from customs ill-fed, ill-rested, disoriented, disheveled, and downright cranky.

You scan the crowd looking for a way out. And, behold! You see a vision. Waiting amid the chaos with a serene smile is a representative from your hotel. You know because she carries a welcoming placard from the Singapore Sheraton Towers with your name on it.

She takes you to a waiting cab, gives you the particulars of the drive—mileage, time, cost, even the name of the driver.

When you arrive at the hotel, your door is swiftly opened. An attendant welcomes you by name. By *name*, mind you. "Welcome to the Sheraton Towers, Mr. Whiteley," he says.

You shake your head in delighted disbelief. The whole arrival, from vision to personal greeting, has transformed you from curmudgeon to human being. And created a fervently dedicated customer.

On the other hand, put yourself in the shoes of a friend of mine who recently checked into a highly recommended hotel in downtown Durham, North Carolina, near Duke University.

My friend had just had one of those days. It's late, he's tired and, yes, cranky. He arrives at the front desk of the hotel, like a thirsty man at an oasis, and breathes a sigh of relief. He languishes on his sore feet for not one, not two, but five interminable minutes while the lone clerk talks on the phone. Finally, she approaches with a map. "What's this for?" he asks. "That's where you park your car," she says, and wanders off to a back room.

Mumbling and grumbling, my friend follows the map, and finds the lot. He figures out the code that opens the lock on the gate. He staggers back with his luggage—with no one watching the lobby, he'd been afraid to leave it while he drove to the lot. Certainly the desk clerk didn't seem interested in watching it for him.

The next day he checked out. He spent the rest of the week—and a good deal of cash—at a nearby Holiday Inn.

The downtown hotel was clean and beautiful. In many respects it was exactly the first-class establishment he'd been led to expect. But my friend didn't take long to decide that its quality was unacceptable—and take his business elsewhere.

"It was a downtown hotel and, obviously, they couldn't help it if their parking lot was inconvenient," he later told me. "What got me—and I guess I never recovered—is that they just didn't seem to give a damn."

■ OUR LIFE-AND-DEATH STRUGGLE

Whether we like it or not, organizations like the company running that hotel aren't likely to survive to the end of this decade. Today the business world is changing radically. It's hardly a news flash that we have entered an era of fierce competition—one in which truly satisfying, even delighting, the customer is absolutely crucial not only to business success but even to business survival.

For a long time after World War II, companies enjoyed a seller's market—a market the likes of which we'll probably never see again. Many businesses lost interest in working for their customers, and yet the customers still came back.

Today some companies continue to prosper. The future of the Singapore Sheraton Towers is secure because I and thousands of other delighted customers will return every time we're in that part of the world. But if the North Carolina hotel doesn't learn to do right by its customers, the customers will find another hotel. If a manufacturer delivers an appliance which I can't figure out how to use or which breaks as soon as I plug it in, I'm unlikely ever to buy from that manufacturer again.

In short, today companies have to do right by the customer *every time*.

Great idea, but how do you make it happen? Many authors recognize the *need* for dramatic changes in the way we do business. But they don't tell how to achieve these changes. Most describe only part of the work necessary to create consistently excellent experiences for customers. And despite all the talk about quality in the past ten years, few organizations have done much research to analyze exactly what behaviors distinguish the companies and work groups that consistently deliver excellence from the much larger number that fail.

■ CREATING THE CUSTOMER-DRIVEN COMPANY

This book is different. It will show you step-by-step how to do right by the customer *all the time*.

In the last few years, a handful of companies have begun vital transformations into a new—or at least hitherto extremely rare—breed of organization: *The customer-driven company*. And I've been associated with a group of people in the Forum Corporation who over the past decade have been closely studying both the expectations of today's customers and the behavior of people in companies that consistently manage to give customers what they want.

This research has produced some very clear results. We've found the companies that deliver what their customers want differ from others in diverse but understandable ways. Perhaps most fundamentally, they provide high quality not according to definitions they've developed on their own but rather *as the customer defines it*. And they achieve that quality in two dimensions—product quality and service quality—each of which calls for different skills and strategies. In fact, the differences between the two dimensions of quality explain a great deal about why so many campaigns for "quality" or "superior customer service" produce disappointing results. Providing one without the other is usually a recipe for failure.

On the other hand, our research on the behavior of people in customer-driven companies shows that the ablest businesspeople have learned an impressive array of reliable practices and techniques over the past decade that have made product and service quality consistently achievable. These methods include both highly disciplined problem-solving practices such as "just-in-time" manufacturing and also "softer," but no less important techniques of good leadership. Most authors concentrate on *either* product quality *or* service quality. They talk about *either* highly disciplined problem-solving tools *or* the demanding human task of leadership. None has shown how to do both.

And yet our research indicates that you can't really gain lasting competitive advantage unless you work at *both* product quality and service quality, and unless you utilize *both* highly disciplined problem-solving tools and good leadership techniques.

But we've also found something else: All the steps the best businesspeople are taking to create customer-driven companies can be summarized in just seven imperatives—the true fundamentals of business success today. This book will tell, in simple language, how to implement each of these imperatives.

That's how this book differs from others on "quality" and "customer service":

1. It's based on careful research into just what behavior and actions result in consistently happy customers.

2. It tells, step-by-step, how to make that behavior and those actions part of the culture of your company or your work group—to achieve lasting competitive advantage.

Businesses have spent many years creating the problems that most of them face today. In the next few pages we'll discuss just where we stand. Then we'll look at the differences between the product-quality and the service-quality dimensions of excellence that businesses have to achieve. We'll conclude this chapter by outlining the seven imperatives that, according to our research, comprise the rudiments of success.

The following chapters will deal, one by one, with how to implement each of the imperatives. Each chapter after this one concludes with a list of "action points" to help you put its lessons to work quickly, and a list of resources—books and articles to which you can turn if you need further help. At the back of this book, moreover, is a "toolkit" that provides special tools and techniques to help you understand and solve the problems of your own organization or work group.

■ QUALITY IN THE LATE TWENTIETH CENTURY: A VERY BRIEF HISTORY

Think back a few years. It wasn't long ago that we viewed foreign competition with arrogance. When Volkswagen sent its Bug against the American auto industry, we laughed. There was a joke about a man knocked off the curb by a dog and then run over by a Volkswagen. "The dog didn't hurt me a bit," he said. "But that tin can tied to its tail nearly killed me."

Some tin can.

We've changed since then, but not enough. By 1980 fat cats in the United States were nervous—and definitely leaner. Americans were out of work, and famous old firms like Ford and RCA were floundering. Internationalization and deregulation had broken down the barriers to competition. What's more, the government was refusing to bail out companies in trouble or pump up the economy with excessive spending.

What to do? Most businesspeople didn't know. They alternated between panic and complacency, and perhaps occasionally found comfort in reading about superefficient Japanese management. When *In Search of Excellence* was published in 1982, it suggested that American companies, too, had lessons to offer, and millions of copies were snapped up. The book was incredibly popular—it was a beacon of hope in a fog of confusion and worry.

Amid this confusion, one collection of strategies began to gain adherents. It was called "the Quality movement." In 1980, NBC television producer Claire Crawford-Mason had discovered that Japan's leading quality award was named for Dr. W. Edwards Deming, an American statistician who was then eighty years old. Deming and J. M. Juran, another American consultant who had helped the Japanese in the 1950's, mixed powerful analytical techniques with exhortations to an outlook Deming called "constancy of purpose." The results of their preachings were sometimes dramatic.

Major United States appliance makers, for instance, cut the number of service calls in the first year after the sale of an appliance by 43 percent, mostly by using techniques like those of Deming and Juran.[2] First National Bank of Chicago not only increased reliability in processing documents such as letters of credit but also cut the average turnaround time in half.

Yet for most companies, even the Quality movement wasn't yet providing the engine that was needed. That school focused on technical perfection—on places where the Deming and Juran techniques could be easily applied—and not on customers, their requirements, and their needs. Wherever the banner "Quality" was unfurled, significant improvements often followed, to be sure. But in most companies the movement attacked

The Customer-Driven Company

only obvious problems that someone *inside* the company considered important, whether or not those were the most serious problems affecting customers. In other words, some companies achieved major short-term gains by picking the low-hanging fruit on the trees in the competitive jungle. But very few had found a route to long-term competitive advantage. Most weren't taking steps that could ensure customers would choose *them* in the future, rather than their hungry competitors.

■ QUALITY AS THE CUSTOMER DEFINES IT

A few companies, however, have done more. The companies that are positioned to earn superior profits today have learned to give us quality as the customer defines it.

Whether it's the Walt Disney Company or Wal-Mart or Honda Motor, they're never tied to somebody's internal definition of "quality." These far-seeing corporations are constantly surveying, testing, and tinkering with their products and services, keeping one goal in mind: to give the customer what the customer wants.

Hideo Sugiura, a Honda executive vice president, offered this view:

> We should not try to sell things just because the market is there, but rather we should seek to create a new market by accurately understanding the potential needs of customers and of society.[3]

Only companies with that kind of commitment to listen and serve can consistently produce delighted customers. And only by delighting customers can you produce robust and growing profits decade after decade.[4]

Just look at what an eighteen-year research effort—the Profit Impact of Market Strategy (PIMS) project—has revealed along this line. In this series of studies, the Strategic Planning Institute in Cambridge, Massachusetts, in cooperation with the Harvard Business School, analyzed 3,000 strategic business units in 450 firms. One finding: When **customer perception of a** business's quality ranked in the top fifth of those in its industry, the company achieved pretax returns on investment, on average, of about 32 percent a year. When quality was perceived as in the bottom 40 percent, return on investment averaged 14 percent less. That's a whopping difference. In fact, high profits correlated better with customer-perceived quality than with market share or any other variable.[5] The companies achieved these returns during an era when it was much easier to make a profit than it is today. You can see why the very existence of the firms that don't please their customers is now threatened.

▪ PRODUCT QUALITY *AND* SERVICE QUALITY

Providing quality as the customer defines it means fully understanding both dimensions of quality: product quality *and* service quality.

The powerful techniques introduced by Deming and Juran deal principally with product quality. If you're a customer, product quality is "What you get." Product quality is usually quantifiable. In a manufacturing company, product quality is the reliability and general excellence of the tangible item that goes out the door. In companies that sell services, product quality consists of the tangible, quantifiable aspects of the service: Does your bank statement carry errors? Is the interior of the airplane clean? Is your hotel room well laid-out and does everything in it work? In most organizations product quality is the province of internally focused, analytical, scientific people.

Now, if product quality is the "What you get" part of the customer's experience, then service quality is the "How you get it" part. If product quality is tangible, service quality can be described as intangible. Thus, it is often harder to measure than product quality. I can calculate fairly easily how often my product broke down in the first year the customer had it. It's much harder to calculate how clear my instruction manuals were, or how friendly my staff was when customers had problems. But that doesn't make those questions any less important.

We used to think of "service quality" principally as what your organization provided when something broke. A company wouldn't think much about service until a customer called with a complaint, and then it would send somebody out to fix the product. It was reactive. And the providers of service quality usually could be found riding around in a panel truck with REPAIR stenciled on the side, or sitting behind a complaint desk.

Today's business environment demands far better, more imaginative management than that. Indeed, organizations today are beginning to understand that service quality, properly understood, can be turned into a highly effective weapon—a way to create and sustain competitive advantage.

Our research uncovered one amazing fact: **Almost 70 percent of the identifiable reasons why customers left typical companies had nothing to do with the product.** In 1988, we surveyed the customers of fourteen major companies in both manufacturing and service industries, serving both business-to-business markets and relationship-oriented consumer

---

■ The Customer-Driven Company: The Research

This book is based on a five-year series of research projects conducted by The Forum Corporation that have sought to define and measure the successful practices that distinguish profitable, customer-driven companies from their less successful peers. Some details from this research appear in the Appendix that begins on page 215.

One study that is particularly relevant in creating customer-driven companies is the Customer Focus Executive Assessment, carried out in 1989. It identified forty characteristics of customer-focused organizations. This research underlies the seven key imperatives identified in this book, and its results can be used to help you analyze how well your organization is doing in vital areas.

A self-test based on these forty characteristics appears on page 220 at the back of this book.

Data from other research by The Forum Corporation and many others are also used throughout the text. The results of a study on leadership for quality are especially important, and are the basis for the book's final chapter.

---

markets such as banking. The reasons customers left them are displayed in Figure 1.

Only 15 percent of the customers switched their business to a competitor because they found a better product, and only 15 percent switched because they found a cheaper product.

What are the prevailing reasons? Poor quality of service. Twenty percent switched because they had experienced too little contact and individual attention. And the largest number—49 percent—said they switched because the attention they did receive was poor in quality.

Here's another example demonstrating why customer-driven companies recognize they must manage quality of service at least as aggressively as quality of product: The Technical Assistance Research Programs Institute (TARP), in Washington, D.C., recently discovered that in two of every three cases in which the customer had a complaint about a product, the problem had nothing to do with the product itself. It had to do with the user. The customer didn't understand how to use the product for the purpose it was designed to serve.[7] Whose fault? Not the customer's; the manufacturer's, of course.

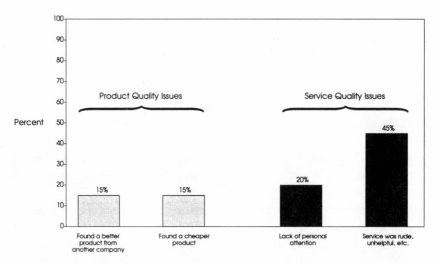

Figure 1   Identifiable Reasons for Switching to a Competitor. Totals do not add to 100% due to rounding. (Source: Forum Corporation[6])

The Customer-Driven Company

Thus, you can outdistance the competition by providing a product that works (the "What you get"), and by providing it in a way that helps the customer make use of it and enjoy it (the "How you get it"). Companies that provide *both* high-quality products *and* high-quality service ultimately win the competitive war.

You create service quality by hiring externally focused people—people who like people—then giving them a vision of service, a knowledge of what the customer needs, and support that lets them do their job. Unfortunately, service quality hasn't been effectively addressed by most of the Quality movement because it is, in general, quite a different creature from product quality. The analytical people in charge of product quality don't usually understand service quality very well. They have trouble fixing what they can't measure objectively.

We can construct a grid (Figure 2) illustrating how much these two very different factors matter. You'll find that some organizations provide excel-

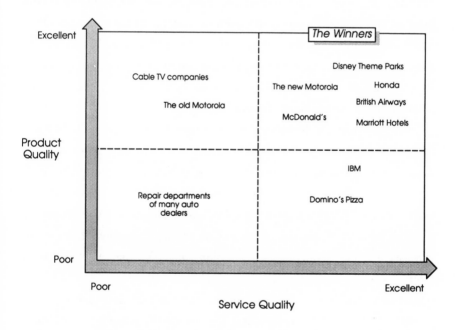

Figure 2   Customer-Focused Quality Grid

lent product quality and mediocre service. Motorola, which was among the first winners of the United States government's Malcolm Baldrige National Quality Award, was once diligent about improving quality in its factories but admittedly lax about training its customer-contact people to improve their relationships—a condition it has taken steps to correct.

On the other hand, IBM was long known for products that were less than brilliant, but for service that was almost legendary.

To be truly successful today, you have to consistently provide both product quality *and* service quality. Figure 2 gives the most fundamental quality picture.

■ WHAT ARE THE WINNERS DOING?

I find a handful of customer-focused companies leave me awed today. Delightfully awed, at that. British Airways, whose flights across the Atlantic definitely were not a high point in my itinerary, now makes me feel like a Lord of the British Empire—and does it with remarkable dependability. Globe Metallurgical, a specialty metals company based in Beverly, Ohio, not only delivers outstanding products but listens to the needs of foundries all over the world and services them so carefully that Globe is now replacing Japanese metal suppliers in Korea.

The Marriott Corporation offers not only excellent physical facilities compared to those of other hotel chains in its price range but also magnificently managed service. Marriott monitors its service quality with intelligent, revealing surveys and motivates employees in a most fundamental way—by treating them well. I've asked dozens of Marriott people at hotels where I've stayed what they think of their company. Their sincere enthusiasm—and loyalty—come bubbling out. Try that at other hotel chains and see what happens.

Even a small group of nonprofit institutions are learning to combine product quality and service quality. The Methodist Hospital in Houston provides exceptional medical care. It tracks its performance against eighty-four measures of medical quality including patient falls, lab-test reporting times, and the frequency with which patients contract infections within the hospital. At the same time the support staff has an admirable quality-assurance program to ensure that wall coverings, air conditioning, beds, lights, and appliances are perfect. Methodist provides services that make a hospital stay resemble a visit to a luxury hotel: inexpensive valet parking for arriving patients and families, complimentary pillows and blankets for visitors who spend the night in a waiting room while doctors treat their loved ones, a bellman to assist patients to their rooms, and admissions personnel who—instead of keeping you hanging around the front desk—will come to your room to complete paperwork.

Will good quality of product and service guarantee continuous success for an organization? No. After all, many other factors have to be considered: the economy, changing taste, and brilliant new competitors. But as Damon Runyon, the writer, said: "The race may not always be to the swift or the fight to the strong, but that's the way to bet."

And surely these organizations and dozens of others we discuss in this book have created the most secure kind of business strength that anyone can build in today's world. They are customer-driven companies, and as long as they continue to be customer driven, the occasional bad years caused by storms in the business world or even misjudgments of the customer's needs will be just temporary setbacks in a long career of success. Guaranteed.

Try a little exercise. Take a minute, and plot your own organization's position on the graph in Figure 3. Where do you live? Then plot one or two of your top competitors. Where are they?

If you're not in the upper right corner, you're highly vulnerable to competitive attack. The rest of this book will show how to get there. And if you are in the upper right corner, guess what? You're a target. Your competition is aiming at you. To win, they will create new forms of product and service quality that no one had ever experienced—extending the product-quality axis and the service-quality axis further than the limits we

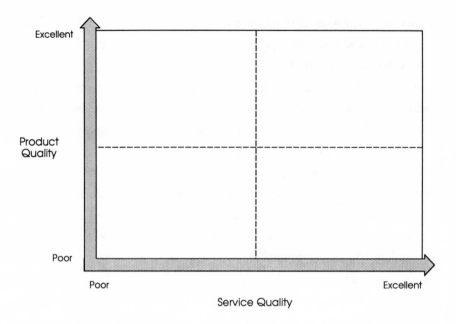

Figure 3   The Quality Grid

know today. It will take all your skill and innovative power to stay ahead of them. But if you know how to execute the real fundamentals of business, you can do it.

■ TWO DIMENSIONS OF QUALITY, SEVEN FUNDAMENTAL IMPERATIVES

Our research shows that the seven imperatives listed here work together to produce a well-integrated organization that can deliver high quality in both product and service. These imperatives have deep roots—in the management ideas America once taught the Japanese and in the customer-driven marketing that helped excellent firms improve their customers' lives in the 1950s. And the best managers are now realizing that these imperatives are fundamental to excellent business in the same sense that practicing scales is fundamental to members of a symphony orchestra, doing calisthenics is fundamental to a dancer, and hitting hundreds of balls at the practice range is fundamental to a professional golfer.

These seven essentials are often the ingredients missing in business. Their absence causes misery for millions of customers, and subsequent downfall for hundreds of firms when competitors arrive and give those customers a choice. Our research shows that these seven essentials are the game plans of today's—and tomorrow's—winners. In the rest of this book you will see how a few companies have mastered some, if not all, of these imperatives, and how they can make you and your organization succeed.

In brief, they are:

1. Create a customer-keeping vision. Nothing does more to transform a company than clear vision. From chairman to switchboard operator, everyone in a dynamic company is committed not just to making money, but to a mission for its customers. Remember how Ray Kroc, the founder of McDonald's, animated the whole organization with the simple vision: "Quality, Service, Cleanliness, Value." Organizations that develop well-defined and widely shared aims wake up their sleepy bureaucracies and truly begin to serve their customers well.

2. Saturate your company with the voice of the customer. Create real intimacy between yourself and your customers. You'll revolutionize your own behavior and you'll remake your competitive position. One example: Binney & Smith, the maker of Crayola crayons and markers, began closely studying the letters it received from customers—mostly glowing testimonials to Crayola crayons' role in children's lives. Managers found that when customers did complain, they usually complained about stains left in clothes. Binney & Smith launched a major research effort, and two years later introduced washable Crayola markers. Marker sales doubled, profits soared, and the company applied the same customer focus to the educational market to produce similar gains through the introduction of fully washable watercolors and finger paints.

3. Go to school on the winners. Great companies cannot hide their way of doing business—and most don't try. Study their methods and philosophies. Those who do study the winners find themselves building commitment to serving their own customers—and learning techniques that will help them discover and eliminate the causes of customers' dissatisfaction. Japanese companies learned to be great after World War II by studying the best Western firms. Companies like Corning Works, Motorola, and Westinghouse made major comebacks partly by studying the best firms in other parts of the world (sometimes even their own joint ventures).

4. Liberate your customer champions. Most employees want to serve their customers well. One of our surveys showed, surprisingly, that the factor most strongly correlated with employees' remaining in a company was simply whether employees thought the organization was providing good service to customers.[8] If they did think so, then turnover was low. Managers must show employees that the company's number-one job is to serve its customers—and that they, the employees, are the key to the entire system.

5. Smash the barriers to customer-winning performance. The more we learn about quality, the more we realize that the systems we've built into our own businesses often inadvertently create barriers to serving customers. For example, invoicing procedures make it easy for people to misrecord what the customer has bought; factory layouts allow pieces of metal to bang together; shipping departments set product delivery policies for their own convenience rather than the customer's. But knowing that things do go "clang" in the night also means that we have enormous opportunities to serve the customer better—by eliminating these glitches. Healthy companies constantly challenge themselves to better understand and improve all the procedures that create value for their customers and to ruthlessly eliminate those that don't. Corning has Quality Improvement Teams in every department. Even the famed Steuben Glass business found it could be improved. Comptroller Danny McNeal led a quality circle that investigated why an unacceptable number of Steuben customers weren't paying their bills within sixty days. The team cut overdue accounts by nearly 40 percent while *increasing* customers' satisfaction simply by reorganizing the billing system. Its finding: most customers will pay their bills promptly if (*big if*) they are correct and understandable. Steuben's now are.

6. **Measure, measure, measure.** In the organizations that are improving most rapidly, people measure almost everything that can tell them what kind of job they're doing for the ultimate judge of their effectiveness, the customer. They analyze their performance not only against their own past and their customers' desires, but also against the performance of the people who, anywhere in the world, are doing a job like theirs best. Aleta Holub, manager of quality assurance at First Chicago Corporation, reports that her organization "benchmarks" how well telephone representatives handle calls by comparing them to customer reps at American Express, for example.

7. **Walk the talk.** Successful managers who carry out these customer-focused principles are creating a new view of leadership. Today, top corporate leaders like Sir Colin Marshall at British Airways, Fred Smith at Federal Express, and Robert Galvin at Motorola have shown what real leaders must do. They personally put the customer first. They promote their companies' visions. They become "students for life," constantly seeking new ways to learn. They believe in and invest in their people. They build customer-focused teams, celebrating successes and encouraging collaboration. And finally, they "lead by example," personifying the organization's purpose.

And, of course, these lessons in leadership apply at every level. Individual professionals and people in charge of small groups can be leaders whose influence permeates their organizations. But they too must become students for life, get in touch with customers, establish direction, communicate clearly, and embody their organization or work group's purpose.

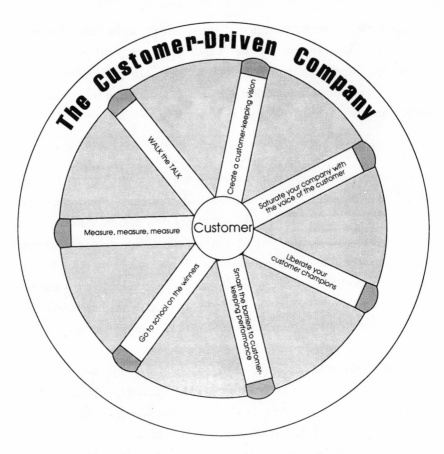

The chart above summarizes the seven fundamentals. Think of this chart as a research-based road map to excellence, growth, and profitability.[9]

You can read this book in either of two ways. Some people will prefer to start from the beginning, work their way to the end, take the self-test on page 220 of the "toolkit" section, and base their work in changing their organizations on the self-test's results.

Other readers may want to go directly to the self-test on page 220. It's a diagnostic tool based on the research described in the Appendix, and The Forum Corporation uses it to help companies understand where they're strong and where they need work. The results of the self-test may suggest you want to read some chapters that appear later in the book before you read earlier chapters.

However you read the book, when you encounter specific problems in your business, consider turning to the sections of the Toolkit on pages 219–293, which provide:

- a section on how to develop a vision,
- specific tools for solving problems that often prevent companies from serving the customer, and
- a guide to selecting the most appropriate tools for solving your own problems.

Current research shows that the seven points are the fundamentals of business success. And yet these essentials still get little attention in today's business schools. Life's lessons have outraced the curricula. But the distilled experience of hundreds of firms shows that these essential ingredients, properly applied, can make your company great. We've carefully chosen the features in this book to enable you to transform your own organization.

■ NOTES

1. *Harvard Business Review* (July/August 1989).
2. Surveys by Baltimore Gas & Electric Co. cited by *Business Week* (June 8, 1987), p. 140.
3. Tetsuo Sakiya, *Honda Motor: The Men, the Management, the Machines* (Tokyo: Kodansha International, 1982), pp. 19–21.
4. In Japan, the phrase "Total Quality" seems to have a clearer meaning than in the West, and the customer seems to have been taken seriously for decades. In his book *What Is Total Quality Control?*, the late Dr. Kaoru Ishikawa said, "To practice quality control is to develop, design, produce and service a quality product which is most economical, most useful, and always satisfactory to the customer." He adds: "We engage in quality control in order to manufacture products with quality which can *satisfy the requirements of the consumers....*We must emphasize *consumer orientation*." (Translated by David J. Lu, Prentice-Hall, 1985. Japanese edition published 1981.)
5. Robert D. Buzzell and Bradley T. Gale, *The PIMS Principles* (New York: Free Press, 1987), pp. 107ff.
6. These data are calculated by further refining data presented in Forum Corporation, *Customer Focus Research Study*, (1988), p. 19. The industries

covered range from retail banking and gasoline retailing to car rental and securities transfer.

7. John Goodman, Arlene Malech, and Colin Adamson, "Don't Fix the Product, Fix the Customer," *Quality Review* (Fall 1988), European edition, pp. 6–11.

8. Forum Corporation, *Customer Focus Research Study* (1988), p. 35. The survey covered 3,300 employees in fourteen companies; the correlation was statistically significant at the level $p = .001$.

9. We don't claim that any of the seven points are new discoveries. But our research has demonstrated, in a way that previous work had not, that these are truly the basics.

We shall build good ships here.
At a profit—if we can.
At a loss—if we must.
But always good ships.
          —*Collis P. Huntington,*
          founder, Newport News
          Shipbuilding and Dry
          Dock Company, 1886

1

Create a
Customer-Keeping Vision

Some companies start out right. When Collis P. Huntington uttered the above words at his shipyard's dedication more than a century ago, he took a clear stand for the customer.

It was not that Huntington was casual about profits. He loved making money and he was highly skilled at it. But he recognized that a customer-oriented vision is essential in driving a company to excellence.

From humble beginnings as a grocer, Huntington had already become fabulously wealthy developing the Central Pacific and Southern Pacific railroads and most of the Chesapeake & Ohio. He was not going soft when he pledged to build good ships "at a loss—if we must." He was recognizing a profoundly important fact: In every era the businesspeople who have taken a clear stand for the customer have also usually become the great winners.

A good vision leads to competitive advantage. Yet today the very idea that a company needs a vision seems to have lost much of its vitality. Even when managers declare that vision matters—even when they insist that it's vital, urgent, important—they seem incapable of creating one, or of making it seem real to the other people in their organizations.

Vision is important for simple, concrete reasons. A vision is the most fundamental impetus in empowering people to serve customers. Without it, employees have little inspiration to do their best. And they lack the unifying ideas that help the people in great organizations join their efforts to achieve seemingly impossible common goals.

This chapter is meant to breathe new life into the whole idea of vision. We'll look at the transforming influence of Huntington's rock-ribbed notions and at the powerful effects of the visions of a few other great business leaders. Then we'll discuss the nuts and bolts of vision, so that you'll have the skills to transform your organization or work group:

- We'll define "vision" in a way that will help you to create one.
- We'll explain the purposes that your vision should be able to achieve.
- And finally we'll examine what you need to do with your vision so that it can inspire the people you work with to new heights of excellence.

■ VISIONARY TRANSFORMATIONS

The effects of Collis P. Huntington's words and vision reached from Newport News, Virginia, across the Pacific. After World War II, American occupation officials began a training course for top management in the Japanese electronics industry by quoting Huntington's statement, then telling the attentive Japanese: "Every business enterprise should have as its very basic policy a simple clear statement, something of this nature."

It was splendid advice, and taken to heart. Most major Japanese companies ended up creating *kaisha hoshin*—statements of "basic policy." And—the crucial step—they worked to drive the whole organization's behavior from them.[1]

The Customer-Driven Company

Huntington's words, however, were mostly forgotten in the United States after World War II. While the Japanese were learning to build their businesses on clear vision, Western companies were neglecting the whole idea. Who needed it, anyway? They focused on profits and financial forecasts, the hard side of business. After all, they reasoned, wasn't that the true purpose of a business, to earn a profit?

This profit-preoccupied view of business dynamics was acted out dramatically in 1968 when the conglomerate Tenneco bought the Newport News shipyard. The new management immediately removed a sixteen-ton monument to Huntington that carried his statement of commitment in large bronze letters, and donated it to a nearby museum. So much, Tenneco seemed to say, for the old fellow's commitment to excellence before profits—a concept as outmoded as the Civil War–era cannons in the same museum. Tenneco was to regret that attitude.

■ Curing the Tylenol Headache: Vision at Johnson & Johnson

Yet you can see the power of a great vision in the behavior of a few United States companies. Often, companies that remembered the vital need for a statement of and commitment to customer-oriented vision became the leaders in their industries. Johnson & Johnson, manufacturer of medical supplies, showed little feeling about money and great interest in people in its statement of values and vision:

> We believe our first responsibility is to doctors, nurses, and patients, to mothers and all others who use our products and services.

This vision—strong and all-embracing—has helped hundreds of Johnson & Johnson divisions, subsidiaries, and offices throughout the world understand that their first obligation is to the customer.

A trial by fire for the company's stated mission was the 1982 Tylenol crisis. A madman (his or her identity has never been discovered) put cyanide in some Tylenol capsules and killed five people. Johnson & Johnson's McNeil Pharmaceuticals subsidiary and the entire Johnson & Johnson organization mobilized within hours to recall the contaminated lot. By the second day of the crisis, when a second incident in California suggested that the poisonings weren't confined to the lot pulled from the market, Johnson & Johnson Chairman James Burke was appearing in news conferences every half hour to keep the public informed.

The company withdrew all Tylenol in the distribution chain, nationwide. As a result, it found two more contaminated pills, prompting Burke to say that "the company believes it has helped to save lives."

Johnson & Johnson could have said, accurately, that the problem was a police matter outside the company's responsibility. But it didn't. It maintained its "first responsibility to doctors, nurses, and patients." Within a week, it had begun redesigning Tylenol packaging with three tamper-resistant barriers. Production of the new packaging began within a month. In the initial struggle, Johnson & Johnson managers gave little thought to the cost.

But here's a point all businesspeople should ponder: After the crisis was past, Wall Street analysts recognized that the fast action had been beneficial to Johnson & Johnson's business. It helped retain the loyalty of millions of customers. Soon after consumers heard about the poisonings, a survey found that half of Tylenol users said they would never buy the product again. But by 1985, the company had recovered almost all of its 35 percent share of the analgesic market. This success was as much the result of a deeply felt vision as of any profit-and-loss business calculation.

Suppose Johnson & Johnson had behaved like the West German auto maker Audi, which faced accusations a few years later that "sudden acceleration" had caused seven deaths and four hundred injuries. Witnesses reported on television's *"60 Minutes"* that Audis rocketed through walls for no apparent reason while people were trying to park them. Audi's own studies indicated the problem was driver error—they showed that cars zoomed out of control most often when they were driven by people with specific physical characteristics (that is, short women). Audi believed that drivers were simply stepping on the accelerator when they meant to step on the brake. From the company's point of view, Audi was no more at fault in the incident than Johnson & Johnson was in the Tylenol poisonings.

But being technically "not at fault" wasn't enough. Sales dropped from 74,000 in 1985 to 26,000 in 1987, and even parking-lot attendants were refusing to touch the car. Audi showed no vision. It simply put out a press release suggesting that driver error was the only reason for the problem. Customers felt insulted and betrayed. They thought Audi was trying to weasel its way out of a responsibility and that they were left with a car widely perceived to be dangerous. Yet Audi showed no apparent compassion, just denial. They should have done much more. Under government pressure, Audi ultimately did modify its foot controls, and the "sudden acceleration" incidents stopped.[2]

You'd expect that customers would trust Johnson & Johnson, the company with a humanitarian vision, buying its products and enabling this firm to consistently achieve returns on equity superior to those of its competitors. It's no surprise, either, that Audi, which had no vision at all, saw its position in the United States marketplace nearly destroyed.

■ A Vision Delivers: Federal Express

Federal Express founder Fred Smith also declared a clear, customer-oriented vision and built a great organization around it. Soon after launching the company, he announced his PEOPLE–SERVICE–PROFIT philosophy:

> We will produce outstanding financial returns by providing totally reliable, competitively superior global air-ground transportation of high priority goods and documents that require rapid, time-sensitive delivery. . . . Production control of each package will be maintained utilizing real time electronic tracking and tracing systems. A complete record of each shipment and delivery will be

presented with our request for payment. We will be helpful, courteous, and professional to each other and the public. We will strive to have a completely satisfied customer at the end of each transaction.

Federal Express has done what the vision promised. It worked with computer companies to invent the technology needed to make its involved communications network and delivery service function efficiently. Federal Express also inspires its couriers to go to extraordinary lengths. One example: When communications equipment in Boston failed during the Labor Day holiday in 1986, the company's trace department in Memphis couldn't confirm the location of a package of blood for Children's Hospital in Boston. Memphis called a Boston courier at home. He made a special trip to the building were Federal Express stored material to be delivered, but found the lock on the fence had been changed: his key didn't work. He scaled the barbed-wire fence, explained the problem to the security guard, found the package, and delivered it.[3] In 1990 Federal Express became the first major service company to win the Malcolm Baldrige National Quality Award. Can you imagine the people of Federal Express creating such an organization without a clear vision?

■ ANATOMY OF A LIVING VISION

What does "vision" mean? And how can you bring vision to your organization?

One good way to define vision is . . .

a vivid picture of an ambitious, desirable future state that is connected to the customer and better in some important way than the current state.

Not all successful visions may fit that definition. But visions that do fit stand a good chance of mobilizing organizations and work groups.

As leaders begin to shape a vision, they ask questions such as:

- What kind of company (or group within the company) do we want to be?
- What will the company be like for our customers and us when we achieve this vision?
- What do we want people to say about us as a result of our work?
- What values are most important to us?
- How does this vision represent the interests of our customers and values that are important to us?
- What place does each person have in this vision of the future?

To wake up a work group, a manager has to state a vision simply and repeat it often. The ideal vision statement is . . .

- clear,
- involving,
- memorable,
- aligned with company values,
- linked to customers' needs,
- seen as a stretch—that is, difficult but not impossible.

Vision is a separate issue from strategy. An organization's strategy is like an architectural blueprint: a clearly drawn design that shows what must be done to achieve success. A vision is like the artist's rendering of a building under construction. A vision excites people as no blueprint—no matter how well drawn—ever can.

■ Of Hearts and Minds: The Functions of a Vision

A vision has two vital functions, and they're more important today than ever before. One is to serve as a source of inspiration. The other is to guide decision making, aligning all the organization's parts so that they work together for a desirable goal.

First, Inspiration . . .

Consider how much today's employees need inspiration. They're often less loyal and more mobile than employees in the past. They're likely to be cynical about their companies, and frankly, they have good reason to be. Overseas competition, deregulation, mergers, takeovers, and the hard times of today's cost-cutting environment have showed them how tenuous their jobs really are.

On the other hand, today's employees seek relevance, involvement, and empowerment. These needs are as genuine as security, housing, and food. A paycheck just won't cover them.

The first function of a vision, then, is to inspire these employees to do their best. A truly integrated and permeating vision energizes people and can resurrect disgruntled, routinized, burned-out employees. It provides true challenge and purpose. It makes each person feel that he or she can make a difference in the world. It becomes a rallying cry for a just cause—their cause.

If your vision is not an impetus to excellence, then it has failed. A major manufacturer of electronic components had as its corporate goal "maximizing the value of the stockholders' investment." Said a member of this company's senior management: "Could you imagine a guy pushing a broom getting excited about 'maximizing the value of the stockholders' investment'?" You couldn't, and the company is now creating a new vision for itself.

Second Function: A Guide for Decision Making

In addition to inspiring people, a well-crafted vision guides them in making decisions. It helps employees cut through complex technical issues, and do only the things that truly add value for the customer. It instructs employees to stay the course during routine decision making. It becomes an oft-referred-to "constitution" during conflict.

As we have seen, having a clearly stated, customer-oriented vision helped Johnson & Johnson make the right choices after the madman struck. Not having such a vision left Audi adrift in a tragic fix that cried out for decisive, compassionate management.

Financial goals give a company nothing distinctive or instructive. Almost any action short of malingering can be argued to advance such goals as increased return on investment or improved market share. Also, financial statistics are *lagging indicators* of a company's success. Many actions that make customers happy won't show their full benefit in market share and profits for months or even years. It's a fact, then: Financial goals don't provide the guidance today's employees need.

A shared picture of what the company should be, on the other hand, fosters independent action, wisdom, empowerment, and willingness to take risks to do the thing that is perceived as right. In a swiftly changing business world, organizations can't afford to wait while decisions are bucked upstairs. All employees need a clear vision as a guide in managing their own actions in a manner consistent with the company's goals.

■ EXCELLENT VISIONS BOTH INSPIRE AND GUIDE

Look at a few fine visions and notice how they have both inspired people and helped them understand what they had to do to serve the customer. Take the "Golden Rule of L. L. Bean":

Sell good merchandise at a reasonable profit, treat your customers like human beings, and they'll always come back for more.

Or the motto that Ford adopted in the 1980s:

Quality is Job 1.

These great visions state the corporate goal simply—and their sponsors communicate their importance by repeating them frequently and applying them to every business situation.

A vision comes to life as it cascades down through an organization. Every department and every individual in the organization should catch, adapt, and "localize" the corporate vision. They should create their own visions of how they will serve the customer. These visions should not only support management's vision but also express their own aspirations in a way that's consistent with their company's overall purpose. What do they want their work group to be?

A legendary example of how vision can transform a job is the janitor working for the Apollo man-on-the-moon project in the 1960s. One day when asked what he was doing, he leaned on his broom and said, "I'm helping us go to the moon."

A three-person benefits department in a small organization came up with this vision:

Benefits are about people. It's not whether you have the forms filled in or whether the checks are written. It's whether people are cared for when they're sick, helped when they're in trouble.

Customer oriented? You bet. This statement is not about administration, processing, or any of the obvious things that benefits people spend their days doing. It's about helping. Isn't that the way we want the people who control our benefits to think?

Workers who build airplanes can be inspired by a vision:

to build parts for airplanes safe enough for our families to fly in.

The manager of a graphics and design department sees her team:

capturing the essence of the company in visual form.

Number crunchers, however talented, rarely wax poetic. And so when a group of financial analysts were sent off to hammer out a vision in a special seminar, their first-draft raison d'être was:

To give people the reports they request on time, sometime in the next three years.

Thud! When their classmates convinced them that this was a minus on the inspiration scale, they took another crack at it. The visioning procedure made them rethink the question: Why do we get up in the morning? They bantered, argued, gave up, went back, and finally produced:

> We are the possessors of information that no one knows about which could represent real competitive advantage. In addition to providing managers with what they ask for, we are going to market the availability of information that they can use to secure competitive advantage.

Still not poetry, but definitely a reason for getting up in the morning.

Something happened in that room that no mere management pep talk could ever have achieved. Their new vision was internally born—it's their own vision, aligned with their corporation's purpose but not produced by a leader's edict. The orders they will follow are their own.

For each individual, the vision has a personal meaning—a meaning perhaps not fully realized in the formal statement, yet completely consistent with it. People arrive at intensely personal and adaptable visions of their own roles. A well-crafted group vision, however, serves as an umbrella under which can be gathered an enormous variety of personal aims consistent with it.

■ MAKING IT REAL

The hard part is imparting the vision to all the people in an organization so that they will share a profound commitment. The leader makes the vision real. He or she

- communicates it constantly,
- establishes challenging, often seemingly impossible concrete goals that are driven by the vision,
- encourages others in the organization to create their own compatible visions for their parts of the business,
- embodies the vision in day-to-day behavior.

British Airways achieved success by stating its vision clearly and then constantly repeating it. Chairman Colin Marshall began the carrier's transformation from inefficient bureaucracy to customer champion with a seminar on its new vision, "Putting People First." All 35,000 employees attended.

A year later, every employee attended another program on the same topic. Marshall understands how important it is to repeat a vision to show you really mean what you say. British Air employees went through the fifth version of the Putting People First program in 1990. They are getting the message.[4]

Over the long run, companies and individual managers achieve their visions by emphasizing different aspects of those visions each year depending on the challenges each succeeding year presents. When Nippon Telephone was privatized, it brought in a former shipbuilding executive, Hisashi Shinto, to convert the organization to a private-sector company. Shinto's vision was a company that would serve customers rather than its own bureaucracy. In his first year he sent every employee a scroll with a four-character message in his own calligraphy: *kyakka shoko*—"reflection on our shortcomings." Later, with the transition to a private company more advanced, he declared a slogan in English—a symbol of international status he wanted the company to adopt. The slogan was:

Best Service

A winner for brevity—and for clarity.

■ When You Take a Stand, You Create a Powerful Network

When a company clearly declares what it stands for and its people share this vision, a powerful network is created—people seeking related goals.

Such a network, we are learning, is the most efficient form of organization anywhere. People on the front lines can make decisions without waiting for approval from above. Yet when faced with a problem, whether it's a manufacturing defect, a complaining customer, or a new sales lead, different members of the network will react similarly.

The United States airline industry in the years since its deregulation in 1978 vividly illustrates both the costs of failing to create a united organization, and the power that can be achieved when a leader embodies a clear vision. Every manager can gain by considering its lessons.

The first years of deregulation produced painful fare wars. They taught the industry that only the efficient would survive, and efficient generally meant big. A big airline could achieve economy of scale, and a big airline could send passengers from one route to another.

Airline managers thought the way to deal with the new world was obvious: Become part of ever-bigger airlines practically overnight, either by buying others or by being bought: an orgy of mergers hit the industry. Texas Air bought Continental, People Express, and Eastern. United bought the Pacific routes of Pan American and also purchased Hilton Hotels and Hertz. Northwest bought Republic. And USAir merged with Piedmont. Crazy times.

Unfortunately, mergers didn't produce new organizations with shared, customer-oriented visions. Instead, they produced confusion. People with different ways of doing things struggled to work together, customers suffered profound indignities, and nobody seemed to care. The promised economy of scale often failed to appear. Airlines were growing, all right, but the efficiency that managers had expected from mergers proved a mirage. Computer glitches, communications failures, and alienated employees left passengers miserable.

One organization, American Airlines, took a different approach. The economics of deregulation shocked American as much as they did anyone else. "If the Wright Brothers were alive today," said American's chief executive, Robert Crandall, in 1980, "Orville would have to lay off Wilbur."[5]

But American thought beyond the illusory benefits of mergers, seeking to understand the full complexity of serving customers in the new, unstable era. It refused to join the merger game.

Instead, the airline laid out a "Growth Plan" that called for no mergers. It sought to create an airline that would be not only bigger, but also better for customers.

Crandall's vision was clear. He sought a carrier that would develop a "comprehensive and profitable route system" by superbly executing all the details involved in serving customers while methodically adding planes and routes. He would hire new crews who could be taught American's way of doing business—and who would initially be paid less than union scale.

Crandall didn't express his vision in a short, inspiring statement: the Growth Plan is three pages long. But American recognized that a workforce with a unified vision could work more efficiently and provide better service than a hodgepodge cobbled together by mergers. American's leaders worked unceasingly to teach every employee to focus on the company's picture of its future.

The airline's goal: to create true economy of scale by orderly growth. Nine top managers of the airline hammered out the Growth Plan in 1982 in the midst of the airline industry's most wrenching recession. American had to choose aggressive growth at a time when it could count on no cash flow from profits.

The Growth Plan is not a plan at all in the usual sense. It can be understood only as a vision, a picture of what the company should be. It appears at the front of the company's employee handbook, and it sets up no numerical objectives at all. It begins:

Growth-plan strategy	American's strategy to deal with the forces of airline deregulation is called the Growth Plan. Important components of the Growth Plan are the development of a comprehensive and profitable route system and the reduction in the difference between our costs and those of the low-cost carriers.
	To sustain and broaden American's established industry leadership, each employee must focus on the Company's Strategic Plan.

Strategic plan   We must maintain a solid business foundation by:

- Operating the safest, best maintained and most reliable fleet in the industry
- Sustaining the momentum of the Growth Plan
- Capturing a premium share of the industry's revenues
- Preserving our reputation for service excellence
- Effectively managing our resources and assets
- Continuously reducing our cost per ASM (average seat mile) and eliminating waste.

American Airlines is a service business. Therefore, we must seek every opportunity to maintain the goodwill of all customers. We must perform and be perceived by the public as the quality leader by:

- Setting high standards and striving to exceed them
- Maintaining our commitment to safety, quality and reliability
- Acknowledging that the success of American Airlines and, therefore, the individual successes of its employees, is dependent on customer satisfaction. . . .

The plan also calls for dignity, respect, "fulfilling and stimulating working conditions," responsive listening, and encouragement of suggestions.

But like any other vision, American's growth plan could have been dismissed by employees as meaningless rhetoric. In fact, many employees initially did so. But ultimately, American's employees learned to perform much as the growth plan said they should.

The reason was that Crandall and other top company leaders constantly worked at reflecting in their own behavior every aspect of the Growth Plan. While other airline presidents were busy trying to buy each other out, Crandall was talking to American's 70,000 employees. The company conducts twenty-five to thirty "President's Conferences" every year throughout the company's 156-city route system. Most employees attend at least one conference each year, and no gathering ends until the questions of every employee present have been answered. The company also sends a daily newswire to all employees. Effective? It seems so: Recent surveys show that employees consider the official pronouncements on the newswire more reliable than the informal company grapevine.

Crandall is a tough boss—probably sometimes too tough. But he has to be considered a great leader. As a result of American's concentration on the basics of organization building, American's people work together unusually well. American has had the best on-time record of any major United States airline and the fewest involuntary "bumpings" caused by overbooking. Frequent flyers consistently rate its service best. Equally important, American's cost per seat-mile has declined since 1982 while those of competitors such as United Airlines and Northwest Airlines have risen. The power of American's vision is visible both in its happy customers *and* on the bottom line.

■ I Take It Back: Newport News Returns to Its Roots

Even if most companies today don't know how to make a vision real for their people, at least we can see a growing recognition that vision matters. Indeed, by the late 1980s even Tenneco, the conglomerate that had removed Newport Shipbuilding's monument to Collis P. Huntington, had seen the light.

Tenneco produced a none-too-spectacular average return on equity of 8.2 percent in the mid-1980s.[6] In 1987, with a dramatic move it hauled the Huntington monument from the museum and put it right back at the entrance to the yard. Newport News Vice President Richard Broad, an up-from-the-ranks executive, speaking to his fellow employees, recalled joining the yard in 1938 when it was building ships that would help win World War II:

> When I was an apprentice, I stood in line every week—along with thousands of other people—behind the main office to get my

cash pay. And with us stood the monument. Every week, we saw those words:

We shall build good ships here.
At a profit—if we can.
At a loss—if we must.
But always good ships.

We all understood that we weren't to take those words literally. We knew that neither Collis Huntington nor Homer Ferguson [the shipyard's first manager] had ever said it was acceptable to lose money. What we did understand was that the motto was an affirmation—a testament—of the Shipyard's continuing and unfailing commitment to quality—past, present and future—and to building nothing less than "good ships" always.[7]

The lesson is simple: Vision makes a difference.

ACTION POINTS

- Review all the statements of purpose and slogans your organization and your work group have used over the past several years. Do any of them convey an appropriately vivid picture of what your organization is supposed to look like and how it is supposed to serve the customer? If not, work with others in your organization to use the "Tools for Developing a Vision" section on page 226 in the toolkit at the back of this book and create a new statement of vision. Someone from outside the immediate work group can frequently be helpful in this project.
- Constantly communicate your vision for your organization to those who work with you and for you. Don't let a day go by without talking about it.
- Help others in the organization create their own compatible visions for their parts of the business.
- Regularly review your own conduct. Are you acting as if you mean what you say? Do your actions embody the vision?

■ RESOURCES

Karl Albrecht and Ron Zemke, *Service America: Doing Business in the New Economy* (Homewood, Ill.: Dow Jones-Irwin, 1985).

Stan Davis, *Future Perfect* (Reading, Mass.: Addison-Wesley, 1987).

Gary Hamel and C. K. Prahalad, "Strategic Intent," *Harvard Business Review* (May/June 1989), pp. 63–76.

Rosabeth Moss Kanter, *The Change Masters* (New York: Simon and Schuster, 1983).

John F. Love, *McDonald's: Behind the Arches* (New York: Bantam, 1986).

Carl Sewell and Paul B. Brown, *Customers for Life* (New York: Doubleday, 1990).

Benjamin Trego and John Zimmerman, *Vision in Action* (New York: Simon & Schuster, 1989).

■ NOTES

1. The story of the Civil Communications Section course is told in Kenneth Hopper, "Creating Japan's New Industrial Management: The Americans as Teachers," *Human Resource Management* (Summer 1982), pp. 13–34, and Robert Chapman Wood, "A Lesson Learned and a Lesson Forgotten," *Forbes*, pp. 70–78.

2. Peter Kuykendall, unpublished paper, February 22, 1990.

3. Bohl, Don Lee, ed., *Close to the Customer: An American Management Association Research Report on Consumer Affairs* (New York, 1987).

4. Naturally, the program has become known at British Air by its initials, PPF. Rank-and-file employees joke that all the repetition suggests the initials really don't stand for "Putting People (the customers) First," but rather for "Pounding People (the employees) Flat." But it may not be a bad idea to pound the message of service into your employees. At British Airways, the program has helped people create their own visions of their jobs and has led them to serve customers well.

5. John Newhouse, *The Sporty Game* (New York: Knopf, 1982), p. 82.

6. This and other return-on-equity data for individual companies in the 1980s come from *Forbes*, "Annual Report on American Industry" (January 8, 1990).

7. *Shipyard Bulletin*, Newport News Shipbuilding (January 1987).

*You just listen to the customers, then act on what
they tell you.*

        —*Charles Lazarus*
          Chief Executive, Toys 'R' Us[1]

2

Saturate Your Company
with the Voice
of the Customer

Donald L. Beaver, Jr., is the founder of a rapidly growing company in
Tipton, Pennsylvania, with a down-to-earth name: New Pig Corporation.
Beaver may sound like a masochist—he says he loves to hear customers
complain. "You should love complaints more than compliments," he says.
"A complaint is somebody letting you know that you haven't satisfied
them yet. They have gold written all over them."

Pay attention to Beaver and others like him, and you'll prosper. New
Pig sells socklike tubular nylon bags filled with ground-up corncobs and
other materials that absorb great quantities of oil, grease, and other gooey
substances. Garages, chemical companies, and practically anyone else who
cleans up the environment, use these "pigs."

Beaver has made a discovery: Every complaint gives him an opportunity
to differentiate his company from others. It helps him find a way to do
something his customers need that competitors aren't doing. Some New Pig

customers complained that the "pigs" were turning to mush when they were exposed to acids and other solvents. Beaver could have said: "Why don't you read the label? The product isn't designed to handle acids." But he didn't do that. He worked with a customer to develop Hazardous Materials Pig, a premium-priced product that handles those chemicals. In response to another complaint, Beaver's company developed Pig Skimmer, a sock that floats on water and absorbs oils.[2]

Too many companies think of their customers as picky, hard-to-please people whose "me-me-me" whining merely indicates they don't appreciate good products. That's a dangerous attitude; research, in fact, shows that it's likely to be absolutely disastrous. Surveys by the Technical Assistance Research Programs, Inc. (TARP) of Washington, D.C., show that in company after company a large share of customers were dissatisfied enough to switch to a competitor. But only 4 percent of dissatisfied customers complain. For every complaint received at company headquarters, twenty-six more customers were unhappy.

Bad news. But worse than that, from 65 to 90 percent of the unhappy but noncomplaining customers would never again buy from the company they felt had wronged them. They would quietly go away, without a word. If you lose seemingly docile, uncomplaining customers in that way, your whole business will slowly erode because you have no idea what went wrong or how to fix it.[3]

The best managers, on the other hand, have discovered an exceedingly valuable tactic. They **invest in complaints**—resolving them. Why? A study for Travelers Insurance showed that persuading people to complain could be, in fact, the best business move a company could make. Only 9 percent of the noncomplainers with a gripe involving $100 or more would buy from the company again. On the other hand, when people *did* complain and their problems were resolved quickly, an impressive 82 percent *would* buy again.[4] And add that benefit to the valuable information about business problems and possible new products that well-run companies gain from complainers.

Professor Jagdish N. Sheth of the University of Southern California estimates it costs five times as much to replace a typical customer as it does to take actions that would have kept the customer in the first place. Companies that can't hear the voice of the customer spend millions on sales and marketing just to replace the customers they're losing. And they don't have a clue as to why their businesses are in trouble.

■ Let Customers' Needs Drive Your Whole Organization

The only right way to run a company—and the most profitable way—is to saturate your company with the customers' voice. Follow this practice and most customers won't end up dissatisfied. You'll hear the voices of both those who are happy and those who are not, and you'll use the information to give an even better experience to your customers. Customer-driven companies create the transforming, company-wide belief that, as Harvard Business School's Theodore Levitt advises:

> Industry is a customer-satisfying process, not a goods-producing process.

Know your customer as you know your own family, satisfy the customer completely, and you'll succeed.

Sound impossible? Don't you believe it.

Anybody can saturate a company with the voice of the customer. The key is that everyone throughout the organization, starting with the leader, must gauge every action against customers' needs, expectations, and wants. If an action isn't adding value for a customer, simply eliminate it.

You can achieve this kind of customer-driven behavior if you accomplish three things. They're difficult, but they're not unachievable for anyone.

- First, carefully target whom you want to be your customers. For the company as a whole, this is generally a top-management decision. And based on the company's vision and top management's understanding of who the company's customers are, individual work groups have to identify their internal customers, the people in the organization to whom they pass the results of their work and whom they have to satisfy so that the organization can satisfy its external customers.

- Second, get to know those customers better than they know themselves. Your entire organization must work to identify what customers need and expect, now and in the future.

- Third, inspire everyone in the organization to measure every action against customers' needs and expectations, and to strive constantly to exceed those expectations.

■ WHO ARE YOUR CUSTOMERS?

Your customers include everyone whose decisions determine whether your organization will prosper. That may be a complex, multilayered group, but you've got to know and serve them all to guarantee prosperity.

Most people in business serve three kinds of customers:

- **Final customers.** People who will use your product or service in daily life and, you hope, be delighted. They're also known as end users.

- **Intermediate customers.** These are often distributors or dealers who make your products and services available to the final customer.

- **Internal customers.** People within your organization who take your work after you've finished with it and carry out the next function on the way toward serving the intermediate and final customers. For an assembler, the internal customer may be the next person on the assembly line. But if you're a corporate leader, the internal customer is often someone who ranks lower than you do. A CEO's internal customer is often his or her secretary, and if the leader fails to prepare directions clearly enough to satisfy the secretary, they're not likely to be executed properly.

For some companies, intermediate customers are worth every bit as much as final customers. A mortgage banker may find the most valued customers are the real-estate brokers who recommend mortgage bankers to the ultimate home buyer. A packaged-food manufacturer such as Campbell's Soup must please the decision makers in supermarkets who allocate shelf space among product lines. Identify your key intermediate customers and make sure they're not being neglected.

Though intermediate customers are vital, it's also possible to over-emphasize them. Don't let your intermediate customers (or your internal customers) distract you from the final customer. If you don't satisfy the home buyer or the consumer who buys the soup off the shelf, you won't succeed. You can even destroy your company. In the 1960s and 1970s, United States automakers generally thought of their dealers as their key customers. This was a myopic outlook. Dealers saw few reasons to complain about manufacturing defects because they made money fixing them. End users weren't so tolerant, however—especially when they discovered higher-quality imports.

For corporate leaders, choosing the customers on whom the company will focus is a major opportunity. Doing it well can lay the foundation for success. Jan Carlzon revolutionized the money-losing Scandinavian Airlines in 1981 by recognizing that business travelers were the key to its comeback. Amid a steep worldwide recession, he saw that businesspeople were the only stable source of revenue on SAS's routes. He declared that SAS's goal was to become "the best airline in the world for the frequent business traveler."

Thus SAS had to understand the business traveler's needs intimately, and it had to inspire each employee to measure every action against those needs. Carlzon communicated his understanding of the business traveler throughout the organization. Then SAS eliminated departments that didn't help meet those needs, including the entire group that promoted vacation flights. The airline taught business-oriented service courses to 12,000 staff members, and paid attention to such details as having olives on all its planes for businesspeople's martinis. The SAS EuroClass, the first "business class," became an industry leader. The company's earnings rose by $80 million when most competitors were posting huge losses.

Choosing whom you should serve is a subtle and demanding challenge, however. Focusing on the business traveler was right for SAS, especially during the 1981 recession. SAS continued to carry tourists, even offering them discount fares. But the airline wisely chose to dedicate most of its energy to serving business fliers.

When Carlzon had run the Swedish domestic airline Linjeflyg before taking over SAS, however, he had then defined his customers differently. Linjeflyg had previously focused too much on business travelers. It flew them in to Stockholm for morning meetings and back home at night, but its planes flew almost empty the rest of the day. Linjeflyg had to be reoriented toward serving two kinds of customers: not only business travelers, who cared mainly about convenient departure times, but also nonbusiness passengers, who needed low fares. A different business called for a different decision. Linjeflyg made a new array of low fares available on off-peak planes, and promoted them so that students, families visiting relatives, and retirees could understand them. The airline prospered.[5]

Choosing the right customers is often the first major challenge a company faces in turning itself around. You've chosen your customers wisely if:

1. The customers have needs you can meet and the means to make sure you're adequately paid for meeting them.

2. Your organization has, or can create, basic strengths that enable it to gain profitable market share by fulfilling those needs better than anyone else.

■ KNOW THY CUSTOMER

After top management defines who your customers will be, the entire organization has to become involved in the never-ending job of understanding those customers. Smart businesspeople understand their customers as well as—indeed, even better than—they understand themselves.

A market-share-building enterprise constantly challenges itself to answer four questions:

1. What *are* our customers' needs and expectations, and which of these needs and expectations matter most to them?

2. How well are we meeting those needs and expectations?

3. How well are our competitors meeting them?

4. How can we go beyond the minimum that will satisfy our customers, to truly delight them?

The most basic step is simple: Ask them how well you're currently serving them. Give them a chance to tell you what they want, where you're failing, where you're succeeding. The companies that do so consistently and act upon their findings, reap massive benefits.

Take Red Lobster, the largest seafood dinner-house chain in the United States. Red Lobster does more than a billion dollars of business a year, and serves nearly 70 million pounds of seafood. Every month it mails a questionnaire to 15,000 customers, asking them to compare Red Lobster with fifteen competitors. If their patrons' taste patterns are changing and their customers want a dish not on the menu, the surveys are designed to show that. Red Lobster restaurants will be serving that dish almost overnight. The company thinks that's the way to create competitive advantage.

For a concrete example of thoroughness in customer communication, look at the Marriott guest questionnaire reprinted on page 65. Marriott no longer uses this format because the hotel chain prefers to mail even more detailed questionnaires to a random sample of customers, but the form shows a good sampling of the array of information a company needs. Marriott could obviously ensure excellence in, say, taking reservations much more easily with this kind of complete feedback from thousands of guests every month than without it.

■ The Spirit of Constant Listening

The best managers—indeed, the best employees of any kind—never cease listening to their customers. In Chapter 6, we'll discuss how to create a rigorous quantitative system for determining what customers want, whether you're now giving it to them, and how each procedure in your organization is contributing to your successes and failures. But people who truly hear the customer's voice begin asking questions long before careful scientific surveys are designed. At every opportunity they ask, "How are we doing?" "How could we do better?"

Psychologist Ellen Langer states that people often get so accustomed to what they do that they can't hear anything new. Listen carefully to some businesspeople talking to customers and you'll find just that. They've heard customers say the service was "Fine" so often that they nod mindlessly, rather than listening aggressively and asking probing questions, when a customer responds with a not-quite-satisfied, "Fine, but. . . ."

They become mentally lazy, like children reacting to these questions, that Langer offers in her *Mindfulness*:

Q. What do we call a tree that has acorns?

A. Oak.

Q. What do we call a funny story?

A. Joke.

Q. What do we call the sound made by a frog?

A. Croak.

Q. What do we call the white of an egg?

A. ?

The correct answer to the last question is, "The white." But nearly everyone gets it wrong. They've become accustomed to words ending in the "oke" sound, and so they answer, "The yolk."[6]

Truly customer-focused businesspeople learn not to react unthinkingly when they talk to customers. They're always thinking of new questions to ask. They ask the questions they think are most important and they also ask open-ended questions such as, "Did anything particularly bother you about your stay?" In that way, customers can paint their own picture of needs. If comments indicate an opportunity for a new product—or trouble with a standard one—the company can act.

They continue this inquisitive, probing approach even when their company has devised highly scientific surveys. The surveys never capture everything—indeed, even the best of them miss some of the most important information. Customers' attitudes frequently change in ways that invalidate the assumptions on which the survey system is based. Or an individual customer may just have a pungent observation to offer that a formal survey system wouldn't capture. The manager who's asking lots of unscientific questions, along with reading reports from a formal survey, will know how the market is changing and how to react. This balance is somewhat akin to the literal reporting that the play-by-play announcer of a basketball game combines with the "color commentary" of a companion announcer, who is usually a former player. Together they are able to provide both facts and texture of the game.

Because quality of service is hard to quantify, companies often fail to learn their customers' opinions about it. But a good guide is structure developed by researchers at Texas A&M University. They determined that a customer's experience of service quality could be described in five dimensions which can be summarized with the acronym RATER:

- **Reliability,** the ability to provide what was promised, dependably and accurately.
- **Assurance,** the knowledge and courtesy of employees, and their ability to convey trust and confidence.
- **Tangibles,** the physical facilities and equipment, and the appearance of personnel.
- **Empathy,** the degree of caring and individual attention provided to customers.
- **Responsiveness,** the willingness to help customers and provide prompt service.[7]

■ Why Don't Customers Complain?

If you think you're doing a good job because your customers aren't complaining, think again. Most people rarely complain when they have a grievance. Why not? They don't complain because . . .

1. They think complaining won't do any good. After all, they know that most employees aren't trained to handle complaints; in most places a complaint will just get you a blank stare. Then why fight City Hall?

2. Complaining is difficult. Think of the last time you wanted to either complain about someone in an organization or compliment their boss. You probably realized that in order to communicate properly, you would have to find out the name of the person, then learn who was his or her supervisor, look up the address, write a letter, and send it. That's a lot of work.

3. People feel awkward or pushy. Most people don't like to complain. They're uncomfortable in that role. On the other hand, a few people are dedicated complainers—they want you to know what they think and they don't much care what you think of them.

One more reason people don't complain is this show-stopper: The competition has gotten so tough and provides so many options, it is literally easier to switch than to complain.

If you know how well you're serving your customers on each of these dimensions and also how well you're providing **value** commensurate to price, you know how well you're doing for your customers—and you have an excellent guide to improvement.

Also ask:

- How your competitors are doing on each dimension.
- Whether your customers have additional ideas about how you can add value to their experience.

Other lists of categories can be used to prepare interview questions and classify data from customers in both manufacturing and service businesses. See page 264 of the Toolkit at the back of the book.

I'll suggest one kind of question to avoid: "Are you satisfied?"

Customers often say, "I'm satisfied" just to avoid hurting your feelings or because they don't believe complaining will do any good. Or because they have no standard against which to compare what you offer. Forum Corporation's research for a bank in California showed that 40 percent of the people who rated service "fair" or "poor" still said they were "satisfied." What would happen if another bank offered them service they considered "excellent"? They'd switch in a minute.

"I'm satisfied" means "I find it acceptable." That's hardly a platform on which to build customer loyalty.

■ Beyond the Basics: Live with Your Customers

To understand our customers, we need to go beyond simple questionnaires. We must get inside our customers' lives, watching them use our product or service, discovering their aspirations and ways of life, their hopes and their fears. In that way we'll position ourselves to respond quickly, even anticipate critical needs long before they themselves recognize them.

Marriott often gets that close to its customers. It continually discovers new ways of communicating. Consider some of its tactics:

- Marriott invests its executives' time in **soliciting and understanding guests' comments.** In 1988, Chairman Bill Marriott, Jr., personally read 10 percent of the 8,000 letters and 2 percent of the 750,000 guest questionnaires the company received each month.[8]
- Marriott conducts dozens of well-designed market-research studies. It mails out **thousands of minutely detailed survey questionnaires** every year. It invites customers to view model hotel rooms; if they don't like the colors, the colors in real hotels will be changed.

- Marriott surveys customers to learn how they react to **specific features offered by Marriott competitors.** If Holiday Inn offers a free continental breakfast, Marriott will know within weeks whether its customers would like one too.[9]
- Marriott has analyzed its guest population and **divided it into segments.** For example, Marriott researchers can predict how each kind of traveler will react to a service: how a new policy in staffing the concierge desk will appeal to businesspeople traveling on their own, how it will appeal to people attending business meetings, and how it will appeal to vacationers.

Smart companies like Marriott are always listening. They realize that this is a struggle that never stops.

The results at Marriott? Occupancy rates have consistently averaged about 10 percent above the industry norm,[10] and the company has earned a return on equity of better than 20 percent over the past ten years. Not a bad reward for an attentive ear.

Many techniques can help a company know its customers' deepest needs. In the next few pages we'll look at a few of them. You can use these to help your company or work group belong to its customers.

■ Invest in Complaints

Smart companies make it easy to complain, and then use the complaints to address the causes behind customers' dissatisfaction. They're putting **toll-free 800 telephone numbers** on packages. A customer can call with a question, complaint, or compliment, twenty-four hours a day.

In some industries, they're actually telephoning customers just to solicit complaints. Nissan telephones each person who buys a new Nissan automobile or brings one in for significant warranty work. The objective is to resolve all dissatisfaction within twenty-four hours.[11]

General Electric has done a remarkable job, too. The owner's manual that comes with any GE appliance lists an 800 number to call for help with any kind of problem. At first, GE worried that it couldn't handle all the calls, and thought they would cost too much. But, no: GE is actually saving money and also increasing sales.

How? First, the 800 number gives GE a way to track how people use its products and how these products might be improved. And it gives the company a chance to manage those complaints. At GE's answer center in Louisville, Kentucky, the specialist who picks up the phone has a personal computer linked to a database with answers to more than 750,000 questions. Let's say you call with a question about a refrigerator. No being put on hold while the people at the center hunt up a refrigerator guru. Instead, the person handling your call brings up facts about your specific model of refrigerator on the PC screen and begins to trouble-shoot.

You'll first be asked, "Is it plugged in?"

"Oh-my-gosh!" you say, "It's not plugged in!" GE has created good will, and avoided a service call and much inconvenience and cost. Listening to the customer can be the cheapest and most profitable way to do business.

Moreover, the 800 line generates requests for product information. With the computer system the operator can identify the nearest high-quality General Electric dealer, tell the caller the dealer's hours of operation, and give specific directions to the store.

The company estimates that GE spends between $2.50 and $4.50 on a typical call, and that the benefits in warranty savings and additional sales are two to three times that. N. Powell Taylor, manager of the Louisville center, says this:

> Most businesses don't understand that customer service is really selling.[12]

■ Get It Direct

Sometimes, particularly with today's technology, you can develop a direct connection with the customer that's even more efficient than an 800 line.

A sale rung up at a Bennetton clothing store anywhere in the world records the type, color, and size of the item. Regional agents report the computerized data to the home office in Trevino, Italy, so rapidly that Trevino can order factories to make more of a hot item—say, to dye undyed gray sweaters in a hot color—and have them in their stores thousands of miles away while they're still the rage.

Technology-based direct feedback especially helps work groups communicate with internal customers who are part of your organization but far away. Subaru of America videotapes problems found in cars when they're unloaded on the docks. For instance, if a car has a dead battery a Subaru staff member explains the problem in English and Japanese on the tape, and the camera operator flashes the serial number. The videotape is duplicated and immediately airfreighted back to Japan. There, the technicians who worked on the cars see the tapes.[13] They won't forget the mistake that's been caught.

■ Romancing the Customer

Once you've decided the customer is Number One, you can discover dozens of ways to draw closer.

• **"Focus Groups" and videotapes of customers.** A focus group is a panel of ordinary customers interviewed about your company and its competitors. These discussions often answer questions you wouldn't think to include in questionnaires. Binney & Smith, maker of Crayola crayons, conducts regular focus groups of children as young as eight years old. To maximize effect on the company, you can videotape customers and replay their comments and their emotions throughout your firm. The focus group probably is most influential when executives sit behind one-way glass and watch customers complain. It's like the Ghost of Christmas Present showing Bob Cratchit's house to Ebenezer Scrooge. Nothing else can change an executive's attitude so dramatically.

- **Executive visits to customers.** Every manager in every company can gain by visiting customers regularly. Hearing a customer directly affects people much more than a thousand carefully crafted reports and charts could. One junior employee for a medium-sized manufacturer of industrial filters launched a revolutionary customer-focus movement in her company by dragging the CEO out to see and hear a few real customers.

- **Employee visits to customers.** Bethlehem Steel sent mill workers to visit customers' factories. The customers were delighted. One factory, about to cancel an order, had complained to Bethlehem's management about flaws in steel plate, but nothing had happened. When a team of mill workers saw the flaws, however, they were practically in tears. The problems were solved overnight.

- **Teaching the front line to listen and communicate.** Make heroes out of low-ranking employees who hear unmet customers' needs and communicate them through the organization. Teach everyone that a key part of their job is hearing the customer and telling their superiors the news—even if it isn't what the leaders want to hear. When Honda Motors first introduced motorcycles in the United States in 1959, its president decreed that it would focus on large bikes like those American makers were then selling. But when salespeople found Californians approaching them and asking where to buy the tiny, inexpensive Supercub scooters that they used, the low-ranking staff were able to reach top management with the message within months and produce a complete change in sales strategy. The result was the Honda motor-scooter boom of the 1960s.[14]

- **Creating the customer's experience for your people.** Find ways of forcing yourself to look at your product from the customer's point of view. Ken Olsen, chairman and founder of Digital Equipment Corporation, once took his executives to the loading dock from which his computers were shipped. He gave each executive a crowbar. Then he asked each one to uncrate a computer and set it up. That replicated the experience of receiving a Digital product.

■ Helping Your Customers to Understand You

Some companies have a penchant for throwing money out of their office windows. They make such impossible promises in their advertising that they're sure to disappoint the buyers. Or they provide instructions so incomprehensible that customers can't possibly enjoy their products as they should. The result: They alienate and ultimately lose customers they easily could have kept. Just by learning to communicate accurately, many companies can greatly improve their long-term prospects.

■ Mean What You Say

Business's failure to communicate is awesome. Research by the Forum Corporation shows customers want reliability more than anything else a business can provide.[a] But businesses don't often give it to them. Companies that genuinely care about their customers often promise what they can't deliver. Think of the Delta Airlines slogan, "Delta Is Ready When You Are." Or how about Holiday Inns: "No Surprises—Guaranteed"? How can anyone deliver on promises like those? Naturally, both of these companies eventually came up with new slogans, but think of the money and goodwill they wasted, promising what they couldn't provide.[b]

Because top management approves of these signals, it's no surprise that employees also mislead customers by promising more than the organization can deliver. Customers call with a need and employees overpromise because they want to show responsiveness. But they're only setting the company up to lose that customer when the organization fails to deliver.

■ Say What You Mean

Similarly, companies routinely publish gibberish in "owner's manuals." One of the happy exceptions is the Buick division of General Motors, whose manual for the Buick LeSabre was described as a "profound change" by *The New York Times*. "A visiting Martian, if taught English and given the manual, could probably change a LeSabre tire," the *Times* commented.[c] The authoritative J. D. Power Initial Quality Survey recently found that the LeSabre, assembled in the same plants as the sometimes maligned cars of other General

Motors divisions, has fewer owner-reported defects in the first ninety days after sale than any other American car. Is there any wonder?

As discussed in the opening section of this book, a study by the Technical Assistance Research Programs Institute in Washington, D.C., found that only one-third of customer dissatisfaction was due to defects in product or service—the manufacturing failures and paperwork errors that quality programs usually try to correct. The remaining two-thirds is **failure to communicate.**[d]

Communicating with customers should be simple:

- First, set realistic expectations.
- Second, acknowledge forthrightly what the product can't do.
- Third, communicate clearly how to use the product or service.

When you do that, you can exceed the expectations you've set. Your customer will be surprised at how well you perform, not angry about surprises discovered after the purchase. And you'll keep that customer.

a. This preference was documented by a Forum Corporation survey of 2,374 customers of fourteen companies. See *Customer Focus Research: Executive Briefing,* The Forum Corporation (April 1988).
b. These examples come from a presentation by A. Parasuraman.
c. Noel Perrin, "Getting to Know Your Synchronized Input Shaft," *New York Times Book Review,* December 24, 1989, p. 18.
d. John Goodman, Arlene Malech, and Colin Adamson, "Don't Fix the Product, Fix the Customer," *Quality Review* (Fall 1988), European edition, pp. 6–11.

- **Customer councils.** Bank of America put its customer councils—groups of customers who meet regularly to advise the company—on stage in the main auditorium at its headquarters. Managers and other bank personnel in the audience fired away with questions for more than an hour. The bank scheduled action-team meetings immediately after the sessions so that managers could begin planning changes to meet the stated needs. And it also sent videotapes of the sessions to branches.

- **Post-purchase assessments.** Survey your customers ninety days, a year, or three years after they buy your product. Ford Motor Company even surveys people who buy its products secondhand as much as five years after manufacture.
- **Competitive product and service questionnaires.** Make special efforts to ask your customers not just whether *you're* meeting their expectations, but also whether *your competitors* are meeting their expectations. Find out where your competitors are beating you.
- **Complaints on videotape.** British Airways set up video complaint booths at Heathrow Airport in London and Kennedy Airport in New York City. Instead of having to find a passenger agent (who, the customer suspects, doesn't want to hear a complaint and is unlikely to do anything about it), a disgruntled passenger simply walked into the booth, pushed a button, and told what went wrong.
- **Formal training in understanding the customer.** As part of their training in "professional customer relations," some employees of Meridian Bancorp in Reading, Pennsylvania, had to fill out deposit slips with Vaseline smeared on their glasses or count money with three fingers on each hand taped together. The bank wanted to give them a better understanding of what older customers with glaucoma or arthritis might be experiencing at the bank.[15]

Finally, pay loving attention to family, friends, and acquaintances who are also your customers. They may tell you what's right and wrong about your product more clearly than any strangers would. Michael Eisner, chief executive of Walt Disney Company, has made his wife and three sons a key "focus group. The Saturday-morning cartoon show *Gummi Bears* was born after Eisner noticed one son passionately loved the sticky, bear-shaped candy with that name.[16]

Notice one more major point: People in your organization can use all these techniques to improve their understanding of their internal customers as well as the final customer.

■ Walk in Your Customers' Shoes

And still, all these methods of getting in touch may not be enough. They can transform your business, yet a competitor can still outdo you by going one step further. To truly serve your customer and win, you have to work hard at getting in touch and *then* you have to think so carefully about what you've heard and seen that you can uncover needs the customer hasn't even thought of.

Nintendo Company has become Japan's most profitable publicly owned firm by selling a game computer that was widely considered technologically obsolete even when it was first introduced. Nintendo's approach differed totally from that of other game and computer manufacturers. Instead of rushing a product to market, it carefully studied prospective customers to learn how to achieve sustained success. Nintendo thought about how customers would use the machine in their homes, and made sure it could offer all the products people would need at prices they could afford. Though the machines had few technical advantages over competing products from Atari and Coleco, or even IBM's PC Jr., Nintendo created the biggest new consumer electronics product since the video recorder. Score one big point for listening *and* thinking.

Dr. Clotaire Rappaille, a French psychologist and president of Archetype Studies, suggests that businesspeople have to understand the "archetypes" in which people think. These are the basic ideas they have about things in their lives, ideas created during their formative years. If a product fits those basic ideas, it can find a place in their lives. If it doesn't, you face the tough job of changing the customers so that they can learn to use the product.

Rappaille tells of a French cheese manufacturer whose product wasn't selling in the United States. Americans who tasted it liked it. But despite instructions to the contrary on the label, Americans put the cheese in the refrigerator, destroying the flavor. In Europe, the company had never faced that problem. What was wrong?

Rapaille helped the manufacturer consider how Americans and Europeans think about cheese. Europeans, they learned, think of cheese as a living thing. One Frenchman said: "I wouldn't put my cat in the refrigerator, why should I do that to my cheese?" But Americans think of cheese as just another commodity, like cereal or orange juice. To them, cheese is dead. As a result, the French company developed for United States customers a "dead" cheese that could be refrigerated. End of problem.

Harvard Business School Professor Theodore Levitt devised another way to think about the customer's needs. Levitt draws the "Total Product Concept" diagram in Figure 1.

Levitt points out that even if your product seems to be a simple commodity—he mentions the chemical benzene—your "product" is much more than just that chemical. The "expected product" customers will buy must include the right delivery schedules, payment terms, and for many, technical support in using the product. The "augmented product" includes

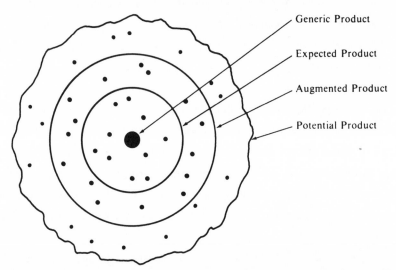

Note: The dots inside each ring represent specific activities or tangible attributes. For example, inside the "Expected Product" are delivery conditions, installation services, postpurchase services, maintenance, spare parts, training, packaging convenience, and the like.

Figure 1   The Total Product Concept

more than the expected product. It includes special "extras" that business-people have created to make their product more attractive, such as special support services for chemicals or the tell-all balance-sheet statements most brokerage firms now offer their customers. As businesspeople offer new augmentations, customers' expectations rise.

The "potential product" is where the gold lies. Beyond asking customers what they need, you must go further to find the latent needs they may not even be aware of. The "potential product" is the opportunity to respond to new needs. Professor Levitt's colleague, E. Raymond Corey, says that

> the product is what the product does, it is the total package of benefits the customer receives when he buys. . . .[17]

Because the "potential product" responds to needs that buyers are unaware they have, a dilemma occurs. If they don't know they have such a need, asking what they need or expect through a questionnaire or a focus group is unlikely to lead you to the potential product. A surer way to discover this gold is to actually observe your customers using your product—or a competitor's product. In that way you can see problems or opportunities of which they are totally unaware.

■ MAKE CUSTOMER NEEDS THE STANDARD FOR SUCCESS

When you're close to your customers, you're on the way to a real competitive advantage. When their needs and expectations become the standards against which your organization measures its efforts and its heart, customers will find their expectations constantly exceeded. They'll experience delight. And they'll respond in a wonderful way—with loyalty.

But to inspire employees to measure their efforts and results against customers' needs and expectations, you have to communicate those needs and expectations and show that you passionately believe they can and must be met. Most companies fail to do that, assuming their people already know what customers need. And yet, front-line employees may fail to understand seemingly obvious points. And when companies fail to discuss customers' needs, employees often think they're not supposed to ask.

Even a little knowledge can make a huge difference. In the early 1980s, welders on the assembly line where Cheverolet Camaros were built kept missing a weld that attached a small bracket. Without the weld, the car's sunshades couldn't be installed. A repair specialist at the end of the line had to take out carpet and trim, weld the bracket in place, and replace carpet and trim. The operation cost GM dearly and added nothing to the car's value. In fact, it substantially reduced the value because the repair rarely got bracket, sunshade, carpet, and trim quite right. As soon as an engineer showed the assembly-line welders the purpose of the bracket and the difficulty of the repair, the problem stopped.[18] Just communicating *why* a job is important and *how* it will serve the customer can often transform an employee's attitude and the quality of the work.

Imagine how much difference it makes when companies communicate customers' needs continually. Early in the summer of 1990 a colleague of mine talked with Darrell Rowe, a manager at the General Electric plant in Lynn, Massachusetts, where they make helicopter engines. He was on his way to Fayetteville, North Carolina, home of Fort Bragg and the 82nd Airborne Division.

"We're getting ready for a Customer Awareness Trip," said Rowe. He explained that 250 people from the Lynn plant, ranging in rank from secretaries and machinists to department managers, would get a chance to see how the products they helped make were being used.

"Many of them have never seen a whole engine, never mind a helicopter," said Rowe. "This experience, we believe, gives our people a real feeling of involvement with the end user—and an even deeper concern about how their product performs." Actually, the 82nd was sent to Saudi Arabia before the Customer Awareness Trip could take place—the value of GE's work for its customers was being tested in a confrontation with a real enemy.

Customers' needs have to be taught, emphasized, and made real constantly. Videotape can create the most vivid effect. I recently saw a twenty-minute videotape at the beginning of a major management meeting at a division of Bell South. It showed customers commenting on the quality of the service they were, or were not, getting. And it had complete credibility and tremendous immediacy. It told the managers more about the customers' points of view than any speech, however stirring, ever could.

Emphasize customer needs everywhere. All meetings of employees are opportunities to teach about the customer. Training sessions, too, are a special opportunity, even if the main subject of the training is as apparently unrelated as computer skills or regulatory compliance procedures. Ryder Systems in Miami, Florida, changed the name of its training organization from Learning Center to Customer Focus Center, for example. Almost every course now emphasizes what the company knows about customers' needs.[19]

■ You and Your Customers: A Partnership

In the manufacturing world, companies use three words to describe the firms from which they buy parts. These names apply equally well outside of manufacturing. Which one applies to your relationships with *your* customers?

- **Vendors** sell a standardized product, often winning business as low bidder. When I buy paper towels or aspirin, I treat the competing manufacturers as vendors. I can't tell the difference among them, and so I buy on price. Vendors are like vending machines. Their job is mechanical, and they're easily replaced. If your customer treats you as a vendor, you don't stand much chance of providing value-added service or gaining reasonable margins.

- **Suppliers** do something unusual, like providing specialized computer chips or pigments for house paint. The customer relies on their expertise.

■ Analyzing Customer Communications

Many organizations receive hundreds of communications from customers every day. These may take the form of requests for repairing broken products, requests for product modifications, or response cards deposited in a comment box.

In well-run organizations, communications result in action. The staff acts on customers' requests, and managers fix problems reported by customers' comment cards. But if nothing more happens in your company, you're missing a great deal of the benefit you could gain from the news your customers are sending you.

A tracking system can monitor feedback from customers and identify recurring problems and priorities for improvement that casual listening to customers would have missed.

To analyze customers' comments:

1. Separate customers' feelings (such as anger and frustration) from the facts. In a journal, record:

- the emotions expressed,
- the expectations expressed,
- their perceptions of what they received, and
- any suggestions for improvement.

2. Group this information using an informal technique such as counting positive and negative comments and classifying them according to the part of your work output that they refer to. (A restaurant manager might want to group positive and negative comments according to whether they refer to the food, the physical premises, the serving staff, and so on.)

3. After you have done this recording for a period (say, a month), try using one of the analytical tools in the toolkit section at the back of this book for a more formal analysis. For example, try:

- a run chart (page 278) reporting the number of various kinds of complaints or compliments in each week;
- a force-field analysis (page 258) with customers' complaints as the driving force and the real causes of the complaints as the restraining force.

The Customer-Driven Company

- stratification (page 288), comparing the average number of complaints on ten nights when one cook is in charge of the kitchen with the average number when another cook is in the kitchen.

A run chart produced for these purposes might look like this:

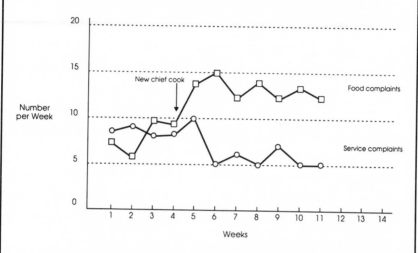

This kind of tracking of customers' comments provides easy access to the information customers are giving you, and clarifies needed improvements.

And therefore the *supplier* can provide better value and gain healthier margins than a *vendor*. The customer also gets more because to maintain its position, a supplier must continually innovate, following its understanding of the customer's needs. Companies often become trusted, profitable suppliers to high price-sensitive customers when they show that they do something unique. Toys 'R' Us has become the supplier to millions of parents who are persuaded that it will always offer excellent selection and value. But a supplier relationship still isn't the best way for organizations to do business. The best way is for seller and buyer to become . . .

• **Partners.** When you become a partner you break down the walls between yourself and your customer. You make a lasting commitment, and you invest in learning everything about your customer and your customer's customers. Your organization can truly become saturated with the customer's voice. You encourage your customer to fully understand what you can do, and you create a "value chain." A chemical company is working on a partnership when the chemical company's researchers study the customer's customers—the apparel makers who buy from a fabric maker or the farmers who buy from a fertilizer company—so that the chemical company can better understand the direct customer's needs.

Partners enjoy trust, a shared vision of the future, free exchange of information, candor, and a long perspective. Not a bad basis from which to discover, create, and provide world-class services or products.

Smart customers want to be partners. Xerox went from 5,000 vendors to 400 parts suppliers because it wanted to build a true partnership with the people making its parts.

Even if your customer doesn't trust you enough to be a partner, it's worthwhile to work toward becoming one. Ford invited customers to drive the Ford Explorer compact truck before its introduction and then suggest improvements. If Ford, through its customer-contact programs, or General Electric, through its answer-center program, treat customers like partners, they'll find a significant number of customers returning that loyalty. Customers recognize real value, and it's a partnership-oriented company, carefully tuned to the customer's voice, which provides real value. A customer who sees you as a sensitive, talented partner will be a customer for life.

Marriott's Casa Marina Resort

OCA 539

1. Please indicate where you obtained this Guest Comment Card.
   - [ ] In my hotel room
   - [ ] With my receipt at check-out (Express Check-Out, Front Desk or Concierge Desk)
   - [ ] Other (please explain)

2. How would you rate our hotel on an *overall* basis?
   - [ ] Excellent [ ] Good [ ] Average [ ] Fair [ ] Poor

3. How was your room reservation made?
   - [ ] This hotel's reservation department
   - [ ] Toll free 800 number
   - [ ] Travel agent
   - [ ] Group reservation card
   - [ ] Housing bureau
   - [ ] Other

4. When you arrived at the hotel, was the information the hotel had concerning your reservation correct? [ ] Yes [ ] No
   If you answered "NO," please check *all* the information which was *not* correct.
   - [ ] Name incorrect
   - [ ] Address incorrect
   - [ ] Arrival date incorrect
   - [ ] Departure date incorrect
   - [ ] Type of room requested not available
   - [ ] Rate incorrect
   - [ ] Party size incorrect
   - [ ] No record of reservation
   - [ ] Other:

5. How would you rate the following?

	Excellent	Good	Average	Fair	Poor
Check-in speed					
Cleanliness and servicing of your room during stay					
Check-out speed					
Value of room for price paid					

   What was the rate *per night* for your hotel room?

6. How would you rate the *attitude* of our staff on an *overall basis*?
   - [ ] Excellent [ ] Good [ ] Average [ ] Fair [ ] Poor
   Please rate the following in terms of their friendly services.

	Excellent	Good	Average	Fair	Poor
Reservation staff					
Front desk clerk					
Bellstaff					
Housekeeping staff					
Telephone operators					
Gift shop staff					
Engineering staff					
Front desk cashier					
Concierge staff:					
Concierge level*					
Hotel lobby*					

   *Not available at all hotels

7. Were *all* of the following in your room in working order? (Engineering items) [ ] Yes [ ] No
   If you indicated "NO," please check *all* which were *not* in working order.
   - [ ] Room air conditioning
   - [ ] Room heating
   - [ ] Bathtub drain
   - [ ] Sink drain
   - [ ] Water temperature
   - [ ] Water pressure
   - [ ] Other bathroom plumbing
   - [ ] Television reception
   - [ ] Television set
   - [ ] Heat lamp
   - [ ] Door latch
   - [ ] Drapes

8. Were *all* of the following in your room in working order? (Other items) [ ] Yes [ ] No
   If you indicated "NO," please check *all* were *not* in working order.
   - [ ] Light bulbs
   - [ ] Clock/Clock radio
   - [ ] Other
   - [ ] Telephone
   - [ ] TV in-room movies

9. Please rate the following which you have used on this visit.
   How would you rate your *dining experience* on an *overall basis*?
   - [ ] Excellent [ ] Good [ ] Average [ ] Fair [ ] Poor

   A. RESTAURANT
   Please indicate name(s) of restaurant(s).

   [ ] Breakfast [ ] Lunch [ ] Dinner

	Excellent	Good	Average	Fair	Poor
Friendly service					
Quality of food					
Menu variety					
Value for price paid					

   Were you seated promptly? [ ] Yes [ ] No
   Was your order taken promptly? [ ] Yes [ ] No
   Was your food served promptly? [ ] Yes [ ] No

   B. ROOM SERVICE (Food and beverage)
   [ ] Breakfast [ ] Lunch [ ] Dinner [ ] Late evening

	Excellent	Good	Average	Fair	Poor
Friendly service					
Telephone order taker					
Server					
Prompt service					
Quality of food					
Menu variety					
Value for price paid					

   C. COCKTAIL LOUNGE
   Please indicate name(s) of lounge(s)
   [ ] 11 am to 5 pm [ ] 5 pm to 8 pm [ ] 8 pm to closing

	Excellent	Good	Average	Fair	Poor
Prompt service					
Friendly service					
Attentive service					
Quality of drinks					
Value for price paid					

   D. BANQUET/CONVENTION EVENT
   Please indicate name of event

If any members of our staff were especially helpful, please let us know who they are and how they were helpful so that we can show them our appreciation.

Name

Position/Comments

10. What was the primary purpose of your visit?
   - [ ] Pleasure [ ] Convention/group meeting/banquet [ ] Business

11. Have you stayed at this hotel previously? [ ] Yes [ ] No

12. *If in the area again*, would you return to this Marriott? [ ] Yes [ ] No
   If you indicated "NO," please check the main reason for not returning.
   - [ ] Other hotels more convenient
   - [ ] Quality of guest rooms
   - [ ] Quality of service
   - [ ] Other
   - [ ] Quality of restaurants
   - [ ] Quality of meeting rooms
   - [ ] Price

13. How many overnight business trips have you made during the past 12 months? _____ trips

14. Are you a member of Marriott's Honored Guest Award Program?
   - [ ] Yes [ ] No

15. Additional comments concerning your stay:

PLEASE PRINT THE FOLLOWING INFORMATION

Arrival date: _____ Departure date: _____

Length of stay: _____ days. Room number: _____

[ ] Mr. [ ] Mrs. [ ] Miss [ ] Ms.

Name

Address

Zip

Business
Telephone: Area code _____ Number _____

THANK YOU VERY MUCH FOR YOUR RESPONSE.
YOUR EVALUATION *WILL* MAKE A DIFFERENCE.

Saturate Your Company with the Voice of the Customer 65

ACTION POINTS

- If in the last twenty-four hours you haven't asked a customer how well you're doing, pick up the phone right now and call one.
- Make sure you've carefully defined the intermediate and final customers your organization will focus on, and that everyone in your organization knows who they are.
- Identify your work group's "internal customers." Are your actions for them serving the final customers?
- Constantly ask your customers how well you're serving them. Assist them in giving you detailed answers that will help every member of your organization to improve. Also ask how well your competitors are serving them, and what you can do for them that your competitors aren't.
- Go beyond questionnaires to find other creative ways of keeping in touch, from visits by executives to real-time electronic monitoring of sales trends. (Use the examples on pages 49 to 56 as a source of ideas).
- Invest in complaints. Help your customers to complain when they're dissatisfied, and use the information to improve your operations.
- Develop a system to communicate throughout your organization the customer's needs and expectations that you're discovering.
- Redirect incentive systems so that you reward behavior serving the customer's interest. Root out incentives that are contrary to the customer's needs.

■ RESOURCES

Jan Carlzon, *Moments of Truth* (New York: Harper & Row, 1987).

Theodore Levitt, *The Marketing Imagination* (New York: Free Press, 1983).

Milind M. Lele and Jagdish N. Sheth, *The Customer Is Key* (New York: John Wiley, 1987).

Richard T. Pascale, "Perspectives on Strategy: The Real Story Behind Honda's Success," *California Management Review* (Spring 1984), pp. 47–71.

Tetsuo Sakiya, *Honda Motor: The Men, the Management, the Machines* (Tokyo: Kodansha International, 1982).

Valarie A. Zeithaml, A. Parasuraman, and Leonard L. Berry, *Delivering Quality Service: Balancing Customers' Perceptions and Expectations* (New York: The Free Press, 1990).

■ NOTES

1. "Ideas for the 1990s," *Fortune* (March 26, 1990), p. 40.
2. "Fending off Copycats," *Venture* (February 1989), pp. 62–64.
3. See Technical Assistance Research Programs, Inc., *Consumer Complaint-Handling in America: Final Report,* White House Office of Consumer Affairs, NTIS PB-263-082, 1980.
4. *Travelers Tribune* (February 1989).
5. Jan Carlzon, *Moments of Truth* (New York: Harper & Row, 1987). Carlzon's story of his successes is an exceptionally clear guide to focusing on the customer.
6. Ellen J. Langer, *Mindfulness* (Reading, Mass.: Addison-Wesley, 1989), pp. 21–22.
7. A. Parasuraman, Valarie Zeithaml, and Leonard L. Berry, "Servqual: A Multiple-Item Scale for Measuring Customer Perceptions of Service Quality," Working Paper of the Marketing Science Institute Research Program, Cambridge, Mass. 1986.
8. *Fortune* (October 10, 1988).
9. *Intrapreneurial Excellence* (August 1986).
10. "How Master Lodger Bill Marriott Prophesied Profit and Prospered," *Fortune* (June 5, 1989), pp. 56–57.
11. "Focusing on the Customer," *Fortune* (June 5, 1989), p. 226.
12. Patricia Sellers, "How to Handle Customers' Gripes," *Fortune* (October 24, 1988).
13. Milind M. Lele, *The Customer Is Key* (New York: John Wiley, 1987).
14. This story is thoroughly documented in Richard T. Pascale, "Perspectives on Strategy: The Real Story Behind Honda's Success," *California Management Review* (Spring 1984), pp. 47–71.
15. Harry Bacas, "Make It Right for the Customer," *Nation's Business* (November 1987), p. 49.
16. Christopher Knowlton, "How Disney Keeps the Magic Going," *Fortune* (December 4, 1989), p. 113.
17. Theodore Levitt, *The Marketing Imagination* (New York: Free Press, 1983), pp. 72–85.
18. Maryann Keller, *Rude Awakening: The Rise, Fall, and Struggle for Recovery of General Motors* (New York: William Morrow, 1989), p. 127.
19. *Customer Service Management Bulletin* (September 25, 1987), no. 115.

*Actually, I'm a good sponge. I absorb ideas and
put them to use.*
                    —*Thomas Alva Edison*[1]

3

Go to School on the Winners

Harley-Davidson executives were flabbergasted in 1982 when they visited the Marysville, Ohio, plant of their strongest competitor, Honda Motor. Honda had seized a 44 percent share of the American heavy motorcycle market. Bikers found Honda's cheaper machines dramatically more reliable than Harley's.

The Harley group expected to learn about the sophisticated technology their competitor was using to beat them. Instead of high technology, however, they found a plant with no computers, no robots, no fancy materials-handling system, and little paperwork. They found a staff of only thirty people besides the 470 production workers, and highly satisfied employees at all levels.

Honda triumphed by rigorously applying common sense, something that Harley could apply in its own factories. Comparing Honda's plant with Harley's, Harley Chairman Vaughn Beals said: "We found it hard to believe we could be that bad, but we were."[2]

Five years later, Harley was an American success story. Its share of United States heavy motorcycle sales rose from 23 to 46 percent. It achieved record earnings of $17.7 million. What had happened? The Marysville visit had led to an attitude revolution. Harley, the symbol of American macho, became a model of humility and teachability. It set out to study the best practices everywhere and adopt them. Harley adopted the best personnel-management systems, the best productivity-improvement methods, and the best quality-control strategies it could find in other organizations. And that turned the company around.

Only 5 percent of Japanese motorcycles coming off the assembly lines failed to pass inspection, Harley learned. But at Harley, the figure was a disheartening and unacceptable 50 to 60 percent. Rejects at Harley due to missing parts alone were several times higher than rejects for all causes in Japan. Sometimes parts that arrived on the Harley assembly line were bad simply because they had stayed in inventory so long that they had rusted, or because minor design changes had made them obsolete.

Why? After many plant tours and much reading, Harley gradually recognized the problems. For instance, the company's inventory control system was at the state of the art by some United States standards. It was designed to control the whole manufacturing process by computer, tracking raw materials and parts as they moved from the receiving dock to a cavernous centralized storeroom, and later passed through the plant on overhead conveyors.

After studying Japanese plants, however, Beals realized that the system actually promoted waste. The Japanese system was based on an uncomplicated secret: Honda and its suppliers produced every kind of part in small batches every day instead of in huge, month-long production runs a few times a year. When parts were made "just in time" for use, companies saved millions of dollars a year in interest charges for carrying their inventories. They eliminated deterioration of parts in storage, saved space, and simplified the operation of the entire plant. When a bad part was discovered, it had been made only a day or two before; the cause was far more likely to be found and corrected.

Harley borrowed the Honda inventory-management system. It adopted employee-involvement systems and statistical quality control. It combined these strategies with its own strengths: ability to communicate with United States bikers and engineer a machine to fit their needs. The result: market share in the domestic heavy motorcycle market surged. Harley became a world-class champion, highly competitive not only in motorcycles but in other markets as well. When the U.S. Army began a Contractor Performance Certification Program to recognize contractors who reduced the government's costs, Harley was the first contractor certified.[3]

■ THE LEARNING COMPANY

Anyone who wants to become a champion needs a dose of humility, which need not come from a crisis like Harley's. But future champions in every field share this common denominator: Because they want to be the best, they're drawn to learn from the masters of their arts. They make ruthlessly honest appraisals of their own goals and limitations. Then they mimic, adapt, transform, and sometimes, if they're very good, surpass the heroes who were the sources of their inspiration.

Name a field of human endeavor and you'll find this method at work. It took Socrates to produce Plato. If you listen closely, you'll hear an echo of Jonathan Winters in Robin Williams. The Beatles learned from Chuck Berry, the Russian ice-hockey team from the Canadians. Matisse absorbed expressionism from Gauguin. And on it goes.

Yet humility and teachability are sadly lacking in businesses today. We have bravado. We have incredible ingenuity. But humility? Hardly. Only a few organizations—such as Harley-Davidson, Emerson Electric, Westinghouse, Xerox, and American Airlines—have understood what can be accomplished by finding out who does what best, how they do it, and then asking: "How can we do better?" Indeed, in the few companies where such wisdom is found, it has usually appeared only after bitter lessons from sharper, faster competitors.

■ A Bank Learns from the Winners

BayBanks, Inc., a $9.5-billion Massachusetts banking company, has made learning from winners a company-wide adventure. Its Legend Maker program sends not managers but outstanding tellers and other front-line people on study missions to institutions like Federal Express, Claridges Hotel in London, and Cookiemania, a superb cookie baker in Chicago.

They interview top managers and discuss strategies, then upon their return present their findings to senior management, including BayBanks President William Crozier. These adventurers and prospectors meet in small groups with Harry Riley, who heads the entire consumer section of the bank, to discuss how their discoveries can be used. The program has been run by Assistant Vice President Mary Jude Dean.

One senior service representative, who normally spends his days discussing loans, new account openings, and service problems with customers in the bank's Malden, Massachusetts branch had created fanatical loyalty among Malden residents. But his voice wouldn't normally be heard much beyond this Boston suburb.

Instead of merely giving him an award or a bonus for serving customers well, however, BayBanks declared him one of the top Legend Makers and sent him as emissary to Smith & Hawken, a Mill Valley, California, garden-equipment direct-sales company known for its careful way of meeting customers' gardening needs.

The service representative brought the fresh perspective of a front-line customer-contact person to the effort to learn from winners. He listened to Smith & Hawken representatives on the telephone and found they "explained the products to customers with as much color and detail as the catalog themselves. The staff is required to have used every piece of equipment in order to better explain its use. . . . Even the software is customer-service oriented.

"When a customer calls to place an order and a customer comment is made," he reported, "the Customer Service Representative can instantly input the comment onto the customer's account screen. . . . At the end of each week, the comments are grouped into a formal report and copies are given to every employee."

He and other Legend Makers told Riley that first-line employees in the bank simply weren't trained in providing the kind of service they found at companies like Smith & Hawken, Federal Express, and Cookiemania. People at BayBanks often lacked the helpfulness that the Legend Makers were finding at the best service companies because they'd simply never been taught how to be that helpful.

It's not that BayBanks didn't train. The bank provided employees with orientation on standards of service, gave them sales training, and taught two-hour courses on its products. But the firm had never taught full-day courses on how to integrate both service and technology to "deliver to the customer something that they want." The Legend Makers urged—and got—a change. BayBanks now has courses that help employees not only fully understand its products, but also how to help customers use them.

"Benchmarking" is a careful search for excellence—taking the absolute best as a standard and trying to surpass it. And not many companies are likely to succeed in the next decade unless they're willing to do just that. The experience of all the companies who've tried it shows that you can enormously improve quality in product, service, and efficiency if you:

- Find teachers—anywhere in the world—people who do a truly excellent job in performing each of the jobs involved in your business.
- Study those teachers, with an open mind and a teachable spirit.
- Never assume you've either found the best or become the best yourself. In other words: Keep searching constantly for better ways.

It's a simple recipe. And a recipe of enormous power.

■ The Roots of Our Haughtiness

Malcolm Forbes said, "The toughest business obstacle to overcome is success." He had a good point. Although benchmarking is almost as old as the wheel, only in the last decade have we seen it used extensively in Western industry. And most of us still don't do it. Our economy grew so smoothly after World War II that most companies took profitable growth for granted. They dismissed new ideas that were "Not Invented Here" and felt their own work was so good they surely couldn't learn much from outsiders. And why should they expect other firms to share the secrets of their success, anyway?

Ironically, these same companies were rapidly giving up their business secrets, and not thinking much about it. "Why worry?" they said, even as hundreds of delegations of Japanese, intent listeners with cameras draped around their necks, took home modernizing business ideas. Western firms had fallen into a downward spiral: Our own inflated self-esteem made us ignore outsiders, and ignoring outsiders kept us from learning their strengths. We therefore blamed market-share losses on "cheap labor" or "unfair" trade tactics or inattentive employees, but rarely on ourselves.

Today, change is finally coming. Still reeling from recent setbacks, companies are going to school on the winners. Improving companies go anywhere to find the masters. Yet sometimes they are right in your back yard. Look at your suppliers, your direct competitors, and even outstanding work groups in your own organization.

■ Success Is Yours for the Taking

Benchmarking is a powerful tool. In my own company, ten officers went on an intensive two-week study mission to Japan. Among other things, we learned that the best companies never stop going to school on the winners. Japanese firms such as Toyota, Nissan, and Komatsu, which have become world leaders, still devote far more time than most Western organizations to learning from everyone they can find.

The strength in learning from the best is enormous. Consider what happened to Robert Toni, former president of Coopervision Cilco, a maker of replacement lenses for human eyes. Toni never went through a crisis like Harley's brush with bankruptcy, but going to school on the winners changed him just as thoroughly. He set out to learn from the best, visiting a dozen companies, including Advanced Cardiovascular Systems, an Eli Lilly subsidiary in San Diego, California. At Advanced Cardiovascular he found thoughtful, quality-oriented operations management that had created world leadership in medical devices such as catheters. And he found his whole approach to business being changed.

"When you see it," Toni says, "you say, 'I can do that.' Because you talk to the people and they're just regular people. And they're very proud of what they've accomplished."

Advanced Cardiovascular's layout was completely different, more efficient than Cilco's. "In our plants we would have a group of lathes, a milling group, a cleaning section, and a polishing section. And of course whenever there was a problem there'd be the question of 'Who's fault was it?'" Toni recalls. American Cardiovascular had brought all operations together in cells, usually laid out in a U-shaped configuration.

"The eight or nine people who make up the cell are totally autonomous," Toni says. "So from the moment you start production till you finish, everything is contained in the cell. These same people are maintaining their own machines, they're problem-solving on their own, they're scheduling production on their own. And they're all multi-skilled. No one is pigeon-holed as, say, a "lathe-operator" or an "inspector."

Back at Coopervision Cilco, Toni changed his whole approach to management. "The impact it had in just the first six months on our defect rates, cycle times, and costs was incredible," he says. In 1986, Coopervision Cilco had a 20 percent return on assets and an 18 percent market share. It was number two in the business of replacement human lenses. Eighteen months later Coopervision Cilco had become number 1, with a 26 percent return on assets and a 22 percent market share.[4]

Of course Toni needed more than just plant tours to transform his company. He emphasizes, however, that theory by itself could never have transformed his organization as study tours did. He adds that the unique characteristics of the medical-products industry had nothing to do with the achievements of American Cardiovascular and Cilco. Other "winners" that inspired Toni included Harley-Davidson and Huffy Corporation, a maker of low-cost bicycles.

"The more you benchmark," says James Sierk, an executive with the Xerox Corporation, "the more you accept that other people are very good at things, even better than you. No one company is good at everything. If you look at the best practices from all over the world, you're going to be an extremely good company."

■ THE NINE-STEP SCHOOL

Plan your learning campaign carefully. Follow this nine-step procedure.

1. Identify your problems. Think carefully about where you're getting beaten. Around 1980, Ford sent study teams to domestic dealerships, both Ford dealers and those that sold Japanese cars, comparing warranty claims. Their finding: Japanese cars had far fewer claims than Fords. That was the first inspiration for Ford visits to Japan, and the information gathered helped them react intelligently as Japanese auto sales soared in the United States.

But you may decide you're behind in very different areas. Do you make mistakes in hiring or promotions? Do your new products suffer too many delays in coming to market? Are you out of touch with your customers? You can make giant steps toward understanding and solving any of these problems by talking to other companies. Do you find your costs are too high and you don't know why? Then you need to look at each part of your production and distribution system and compare it with similar efforts in other organizations.

2. Choose organizations that are solving the problems you face. Find several organizations that are solving the problems you've identified. In your search, talk to suppliers, customers, trade associations, consultants—and don't overlook customers. Don't even be afraid to call up competitors and say something along this line: "We both have problems. I admire the work you're doing. Let's share solutions."

Be sure to visit companies in industries you know little about. Then, instead of focusing on technical details, you're more likely to appreciate the management systems—and it's those you care about.

You may be surprised at your reception. Such attention is flattering, and these days sharing is encouraged. Winners of the U.S. government's Malcolm Baldrige National Quality Award, for example, must, as a condition of the award, help other companies learn from their successes. The winners include Motorola, Xerox, Milliken & Company, Federal Express, Cadillac Motor Car Division of General Motors, and IBM's Rochester, New York, plant.

3. Develop specific objectives before each visit. To use your time well, study your problems before you visit another company. Develop ideas about what you're doing wrong and what others may be doing right. The Belleview, Washington, consulting firm DeltaPoint, which has organized highly successful plant tours like Robert Toni's, requires chief executives to read eight books (most of which are listed in the Resource sections at the end of the chapters of this book) and take an eighty-question test on what they've read before it will lead them on visits to high-performing companies.

Before you visit, prepare a list of questions you want answered. Let your hosts know what you want to learn, and ask them if they'd like to keep any subjects off-limits. If you want to take photographs, ask whether that will be okay. Productive visits always produce surprises—answers to questions you never thought to ask. Clear objectives will help you ask useful questions. And a review of your original objectives will help you understand afterward just how your visit has changed your understanding.

4. Make the visit. Travel with several others from your organization when you visit another company. Then you can compare notes, and you'll have more credibility when reporting surprising findings to the rest of your organization.

On visiting day:

- Take a nice gift for your host. You're getting something valuable from him or her. Show that you appreciate it.
- Focus on interests you share with your host. Tell about your own struggles. Your host probably hopes to learn from your visit, and discussing your own problems may prompt helpful comments from the host.
- Carry a written list of questions, and designate members of your party to ask them. Keep your eye on the clock, so that you won't run out of time before you've dealt with indispensable points.
- Ask about *everything* that's relevant to your needs. Your host will let you know what's off limits.
- Praise and appreciate what you see, and avoid criticizing aspects of your host's company that you believe you could improve. Try to elicit the information you want without giving the idea that you know more than the host.

Finally, don't focus so intensely on the problems or techniques you came to study that you miss seemingly unrelated ideas that could help your firm. During my own company's recent study mission to Japan, we were all amazed at how extensively every company we visited used "visual management"—charts, graphs, signs, even pictures of individual workers—to punch home goals, quotas, progress, and, yes, shortfalls and problems, too.

Effective? Profoundly so. We were looking in other directions, but we could not help but be impressed. We brought another idea home with us, and now the walls of our company are covered with charts and graphs.

5. Debrief. A study trip to another firm will almost certainly produce a welter of confusing ideas. You'll never reap all the potential benefits unless you take time to sort them out. Schedule time for a debriefing session. Set it up before you leave for your visit—or it will never happen. When you return, debriefing will be forgotten in the press of today's problems if it's not scheduled in advance. Also, scheduling a debriefing before the trip is a strong suggestion to colleagues and subordinates that you mean business. This is no casual stroll.

Combining three types of debriefing works best. First, a quick team debriefing (it can start as soon as you leave your host) captures and discusses first impressions. Second, a formal debriefing focuses on what was seen and learned, what discoveries can help your company. And, third, each team member makes a written report.

6. Convert learning to action. After you've decided just what you've learned and what practices you'd like to adopt, get going. Create action teams in your own organization. Set clear goals and establish standards you can use to measure success. How will we know if we've really absorbed and are using this knowledge—unless we measure?

7. Spread the learning throughout your organization. Identify critical practices your organization should be adopting—not just in the few areas where you can cause an action team to begin work, but everywhere. Then package what you've learned and make it accessible to your whole organization. Make speeches, give seminars, create a videotape describing what you've learned. Repeat your message constantly to show that you believe the organization really can change in response to it.

8. Show the winners how much good has been done. Don't forget the company that helped you. Tell the people there what you think you learned. Then later, tell them the results in your organization and any difficulties you've encountered. By keeping in touch, you'll help the winners understand their own processes even better, perpetuating their interest in helping others. And you'll maintain communication so that as new problems appear you can return for guidance.

9. Repeat the cycle. As you begin to see results—or as you figure out why results haven't been forthcoming—it's time to identify problems all over again and seek contact with more organizations that can help you. You've now entered the "continuous improvement loop."

"Life is difficult," said M. Scott Peck in the opening lines of his best-selling *The Road Less Traveled*. Well, achieving success in the business world is, too—but it's obviously worth the effort.

■ Reaching Out to the Very Best

Reaching out to a few good teachers is only the first part of a benchmarking effort. As you learn from more and more organizations, you'll find just who *is* world class in each function of your business.

Benchmark the performance of every function in your business against the best performance of those functions anywhere. Engineers creating the Ford Taurus set out to make it "best in class" in more than 400 design elements. By Ford's own reckoning, they achieved that goal for 77 percent of the elements. But Ford's benchmarking efforts are finding plenty of work that still needs to be done. Japanese firms no longer design cars *better* than Ford, but they know how to design and produce excellent cars in less time.

At Xerox's Document Processing business, all employees know they must stand ready to compare themselves with people doing similar jobs anywhere in the world. There's not much similarity between duck boots and copiers, and yet Xerox has reorganized warehouse order-picking following a study of an L. L. Bean warehouse. Xerox also compares the speed with which the results of laboratory research appear in its products with achievements of IBM, Bell Laboratories, and Canon.[5]

When you become a winner yourself, you go to school on the winners all the more avidly. Stew Leonard's in Norwalk, Connecticut, one of the best-run food stores in the world, owns a bus. Why would a supermarket need a bus? Top managers regularly take groups of twelve employees on "One Idea Club" field trips. They travel to other supermarkets, sometimes as far as 400 miles away. Each employee is expected to find one area in which the other store out-does Stew Leonard's and suggest how Leonard's can do as well—or better.

Marriott sends employees to stay in other hotels. They check on things (benchmarking against Marriott standards) like brands of soap and shampoo, special services, and registration procedures.

And benchmarking often leads to one of the most effective transforming steps any organization can take: establishing an apparently impossible, but actually achievable, goal. Motorola in 1981 set the seemingly difficult goal of improving its statistical measures of quality tenfold in a five-year period. And they achieved it by the end of 1983—two years ahead of schedule. Then the Motorola people toured some of the best companies in Japan, and they learned how reliable manufacturing processes could become. The result was new targets that were vastly higher than the old ones.

"We now realize that you have to have dramatic, impossible goals," says Paul Noakes, Motorola's vice president for external quality. "We'd been going along at 15 percent a year. If you're doing that and you set a goal of 20 percent, people will sweat a little and accomplish the goal without really changing the way they do business. But if you say you're going to achieve a tenfold improvement, then they realize that they have to do something really different."

■ Anyone Can Do It

Anyone can find winners and learn to be a winner by studying them. Perhaps you'll start with your best suppliers and your most impressive customers. When First Chicago Corporation launched a major quality initiative, the firm realized that its customers and suppliers included many organizations known for excellence—3 M, IBM, Westinghouse, Ford—and went to them first for help. Some companies even go to school on the winners by learning from their own joint ventures in Japan.

Smaller firms can start by talking to suppliers such as Federal Express or Xerox. Even government agencies can learn from the winners, seeking out the best-run nonprofit organizations or the most admired businesses in their communities. Larry Nyland, superintendent of the Pasco, Washington, school district, learned about team management from Ford and about the organization of new projects from General Electric. Pasco often sends teachers and administrators to study programs in other school districts. Has all this extra effort helped? Apparently: Pasco has received national acclaim for producing significantly rising test scores in one of the poorest school districts in the state.

Most outstanding companies are glad to help others. Whitman Corporation, the $4-billion maker of products such as Whitman's chocolates, Progresso soups, Midas mufflers, and Hassman commercial refrigeration systems, launched a quality effort by choosing twenty-four United States companies it considered quality leaders and sending a team to each.

Every one of the twenty-four welcomed Whitman managers—even Hershey, whose chocolates compete directly with Whitman.[6] "They recognized that we are in a global battle," says William Naumann, vice president for quality with Whitman. "Here at Whitman, we're in the soup business, and it's interesting to note that the Japanese have taken a significant part of that market" with the introduction of instant Ramen noodles, a hit in U.S. supermarkets.

Your competitors won't help? Not a big deal. Just identify which aspects of your organization seem to need improving. Then go on to noncompeting companies you feel handle those activities well. Do telephone customer-service representatives foul up your customers' orders? Talk to companies like L. L. Bean or Marriott. Does your factory spend too much or produce too many defects? Even if you don't make motorcycles or copiers or glassware, you can probably learn from Harley-Davidson or Xerox or Corning.

■ The Best Companies Are Those Who Know They Need to Improve

The key to learning is finding teachers who do an excellent job of performing the tasks *your* business depends on. Study those teachers hard. And always keep looking for even better ways of doing your jobs.

Studying "the best" can, in fact, drive your entire business. It can put an end to "that's-the-way-we've-always-done-it" thinking. It can show skeptics that change means a better life for everyone. Companies who go to school on the winners serve customers far better than their competitors can.

■ Learning from the Losers

My wife, Sharon Cavanaugh, owns a rapidly growing company called Peacock Papers that manufactures contemporary party, gift, and apparel items. Her product line features such bold, upbeat, and clever (we think) messages as,

> Did anyone tell you today? You're terrific!
>
> Very modern, Very Italian & Very good
>
> A peacock that sits on its tailfeathers
> is just another turkey.

She was Ernst and Young/*Inc.* magazine's regional Entrepreneur of the Year in 1987. Together, we operate a small retail store in Boston's Faneuil Hall marketplace that sells Peacock Papers products. And we've done our best to go to school on the winners in retailing—stores like the The Gap, Nordstrom's, and Crate & Barrel.

But we've also found another technique that affects our business dramatically: Getting the people who work with us at Peacock Papers to look closely at their own experiences as customers.

We believe our store's main purpose is to make people who visit us happy—whether they buy anything or not. We have three resources with which to achieve this end: the products, the premises (the layout of the store itself), and our people.

People are the most important. To drive this point home, we gave each Peacock Papers person some cash and said: "Go out into the marketplace here and spend the money. You can buy anything you want, and whatever you buy, you can keep. Go to at least five stores. Notice, as you walk out of each store, whether you feel like smiling or like frowning."

When they came back we had them complete a little form on which they could list happy or unhappy experiences they'd had with products, premises, or people.

Some of our people listed happy experiences with products, premises, *and* people. Some listed unhappy experiences with products and premises. But when it came to unhappy experiences with people, their reactions were all extremely strong. Every employee filled up the space on the sheet designated for unhappy experiences with people working in stores. Some typical comments, as I recall, were: "She was chewing gum." "I was ignored." "She was talking on the phone while checking me out." "He was obviously bored and didn't want to be bothered."

Did it help us to have our people focus on what it's like on the other side of the counter? We think so. Business is good, and salespeople in our store go out of their way to be helpful, courteous, and friendly. This is hardly a scientific measurement, but I haven't seen many customers leaving with frowns, either.

ACTION POINTS

- Immediately start seeking teachers who can show you how to do your jobs better.
- Follow the nine-point study plan that begins on page 76 of this chapter.
- Rigorously benchmark everything you do against the best people in the world who are doing those jobs.
- Never assume you've "made it." Keep looking for people better than you.

■ RESOURCES

Peter C. Reid, *Well Made in America: Lessons from Harley-Davidson on Being the Best* (New York: McGraw-Hill, 1990).

Robert C. Camp, *Benchmarking: The Search for Industry Best Practices That Lead to Superior Performance* (Milwaukee, Wis.: American Society for Quality Control/Quality Press, 1989).

Gary Jacobson and John Hillkirk, *Xerox: American Samurai* (New York: Collier, 1986).

Ira Magaziner and Mark Patinkin, *The Silent War* (New York: Random House, 1989).

■ NOTES

1. This quotation is sometimes used by Xerox chief executive David Kearns in speeches. Gary Jacobson and John Hillkirk, *Xerox: American Samurai* (New York: Collier, 1986), p. 233.

2. Peter C. Reid, *Well Made in America* (New York: McGraw-Hill, 1990), pp. 13–19.

3. Reid, *p. 133.*

4. Coopervision took advantage of the strength of Cilco to sell it to Nestlé Ltd. for an undisclosed price eighteen months after Toni began his management changes. Though Nestlé continued the program, Toni left the company. He recently became president of the former leader of the industry, Johnson & Johnson's Iolab.

5. Jacobson and Hillkirk, pp. 229–232.

6. The companies they visited were American Telephone & Telegraph, Bank One, Campbell Soup, Conagra, Corning, Federal Express, First Chicago Corp., Florida Power & Light, Globe Metallurgical, GTE, Hewlett-Packard, H.J. Heinz, International Business Machines, McDonnell Douglas, McDonald's, Milliken & Co., Motorola, Nestlé USA, Procter & Gamble, Tennant, Inc., Texas Instruments, Minnesota Mining & Manufacturing, Weyerhaueser, and Xerox.

Customer satisfaction is first and foremost a
state of mind and action . . . an every minute of
every day obsession.
        —Charles M. Cawley,
          President of MBNA
          America, Inc.

4

Liberate Your Customer Champions

Recently I arrived at the front desk of a prestigious Chicago hotel to check in. While the clerk looked for my reservation, I noticed hotel staff members setting up microphones and speakers nearby. They appeared to be getting ready for some kind of musical performance. A frustrating dialogue ensued.

"What's going on over there?" I asked the clerk. Hoping for a respite from an exhausting business trip, I wanted to know when the concert would begin.

"I don't know," he said as he went about his business.

"Who would know?" I pressed.

"I don't know," came the reply. "Maybe the assistant manager."

The clerk continued methodically punching keys on his computer terminal. He showed no interest in finding the assistant manager or anyone else knowledgeable about the day's schedule of events. I was giving up on the concert, but it seemed bizarre for the hotel to plan a special event and then neglect to inform the staff about it.

After completing the perfunctory check-in, the clerk turned away from the computer, handed me my key, looked me square in the eye, and said in a sincere voice: "Thank you very much Mr. Whiteley. My name is John. If there's anything I can do to help you, please let me know."

Incredulous, I looked at him and said as politely as I could: "John, I just asked you for your help. I'd like to know what they're setting up for over there."

Then, and only then, did the light go on for John. I was right—it was a concert organized explicitly for the enjoyment of the hotel's customers.

John had obviously received training. Otherwise, how would he know he was supposed to say, "If there's anything I can do to help you, please let me know"? But the training taught John to process customers, not to please them. John processed me flawlessly by some standards—much as a machine processes data, sausage meat, or cheese. But unfortunately, John's manager hadn't helped him understand the distinction between his job of following a script and his job of creating a happy customer. As a result, although the check-in procedure was technically perfect, the customer was not well served.

Contrast John's unhelpful behavior, which is unfortunately quite common, with they typical behavior of staff members in a company that knows how to manage people's efforts: MBNA America, the fourth largest bank credit card issuer in the United States and a finalist for (but not a winner of) the Malcolm Baldrige National Quality Award.

The MBNA organization teems with knowledgeable service people who are ready to go above and beyond their job description to take care of customers. When colleagues of mine have been connected to the wrong office at MBNA, people have taken the time to figure out where they should have called, and provided the correct phone number. In more essential matters too, MBNA delivers superior service. It approves credit-line increases in one hour instead of the eight-day industry average, and it replaces lost cards in three days instead of the industry average of nine.

Most important, MBNA accomplishes a remarkable feat: It helps its people serve customers well and enjoy doing it. MBNA is located in Newark, Delaware. Many banks have established credit-card organizations in that area to take advantage of favorable Delaware banking laws. And in the area, the average bank loses 25 percent of its people every year. That means monstrous recruiting and training costs. It means that hundreds of people, new to the job, lack the knowledge to properly serve the customer.

In the average twelve-month period MBNA, in contrast, loses only about 7.5 percent of its staff. MBNA people simply like what they're doing.

MBNA America is an unusual organization. These are troubled times for banks, and MBNA's own parent company, Baltimore-based MNC Financial, has suffered along with the rest of the industry (The letters MBNA originally stood for "Maryland Bank, National Association). MNC Financial's losses are so great, in fact, that as this is written it has had to put MBNA up for sale. But MBNA is almost universally recognized—by hard-nosed financial analysts on Wall Street as well as by its customers—as a jewel. MBNA has grown about three times as fast as the credit-card industry as a whole over the past six years.

MBNA offers lessons to all managers. Charles Cawley, the president and CEO, has developed an unusually clear understanding of how to manage people. Cawley's ideas are so direct and widely applicable that I'd like to quote portions of a speech he delivered to the National Quality Forum in 1989. These are MBNA's nine critical beliefs, and they form an excellent basis for managing people almost anywhere:[1]

- First, **set an unswerving goal of putting Customer Satisfaction above all other objectives**. . . .

- Second, **tell everyone about it**. . . . Everyone has to believe that there is an *absolute obsession* with quality—with Customer Satisfaction.

Don't be embarrassed. If you believe something is, *it is*. . . . At MBNA we put it in front of ourselves in every way possible. We think about it, we talk about it, we even paint it on the walls. It's printed on the envelopes our paychecks come in; we have parties to celebrate it. . . . We can and do trace our financial success directly to it.

- Third, **measure Customer Satisfaction daily—post the results and reward everyone when it is achieved.** . . . What gets attended to gets done. . . .

- Fourth, **hire people who like other people.** Hire and retain people who will accept and enthusiastically carry out the things necessary to satisfy the Customer. At MBNA, senior executives interview each candidate for every job in the company—whether marketing director or carpenter in the maintenance shop. Their objective: to ensure that the candidate is *a person who likes other people.* If you hire people who like people, one of the bottom-line results is that a large portion of your new hires will be referrals by people who already work in your company. At MBNA, 80 percent of the people we hire are recommended by current MBNA People. And they stay longer and perform better. . . . When you hire people who like people, your Customers will notice. . . .

- Fifth, **let them know what to expect and what's expected of them.** Every successful candidate for employment at MBNA reads and signs a list of precepts which clearly spells out expectations. Among a list of ten are these:

  - The People of MBNA can expect to be judged individually by the quality and consistency of their effort, enthusiasm, honesty, and results; and,
  - Above all, the People of MBNA can expect to work hard in an environment absolutely committed to excellence; and excellence means quality; and quality means Customer Satisfaction.

In turn, what is expected of the People? Just this—the People of MBNA are expected to treat the Customer as they would like to be treated; to put the Customer first every day; and to mean it.

- Sixth, **educate people from the time they're hired.** During the first month, every new MBNA person attends the Customer College. . . . The overriding focus of the college is to ensure that all new People understand that the very existence of MBNA rests on Customer satisfaction.

- Seventh, **create an environment that makes people feel good and supports their enthusiastic pursuit of Customer Satisfaction.** . . . It's everyday things like good food in the cafe, convenient parking, clean rest rooms, sharp pencils, spotless work stations—things that make people feel good. And when people feel good, they transmit that feeling to their job—they satisfy the Customer.

- Eighth, **treat People like Customers.** You may have noticed during my remarks that the term "employee" has never come up. We don't have "employees" at MBNA. In fact, "employee" is a *non*functioning word at our company. It reduces people to a category and carries with it undertones of ownership. *People* work at MBNA. . . . We do not manage these People. Rather, we manage their *efforts.* The People of MBNA are treated as Customers. Treating the company's People like Customers sets a clear example. . . . At MBNA, we are all each other's Customers.

- Finally, **think of Yourself as a Customer.** People who like People find it natural to take the Customer's point of view. They easily think of themselves as Customers. If you hire the right kind of People, then it's enough to say to them, "Think of yourself as a Customer" and they'll do the rest.

This states powerfully the truths that management experts have learned over the past few decades about inspiring and motivating people.

And because MBNA practices it, its credit-card business is extremely profitable. Today, MBNA has nearly 8 million cards outstanding. MBNA's rapid growth is among the most impressive achievements of any company in the service industry.

MBNA pioneered "affinity card" marketing—the issuing of credit cards in cooperation with such groups as the American Dental Association and local medical associations. Affinity-card holders get a group of special services, including hefty credit lines, free insurance benefits, and toll-free numbers they can call for help when traveling.

Non-profit organizations such as Ducks Unlimited, a wetlands preservation group, can even arrange to receive a donation each time a member uses one of their cards. MBNA provides affinity cards for 1,300 groups.

The affinity card was a brilliant business idea, and MBNA's customers (affluent and excellent credit risks) use their cards twice as often as the average bank credit-card holder. What's more, MBNA's credit losses are 40 percent below the industry average.

But MBNA's invention of affinity cards wouldn't have created sustainable success if it weren't for MBNA's way of hiring and developing its people. In today's economy, companies must not only innovate, they must also deliver their innovations to customers with exceptional quality of product and service. If others move faster and better, you'll lose your advantage. Many other banks now compete aggressively with MBNA in offering card programs.

In this chapter I'll dig deeper into Cawley's people-management principles. They can help anyone create the same kind of happy, profitable organization as MBNA has built. We've covered the first two principles—an unswerving goal of customer satisfaction and constant communication of that idea—in earlier chapters. We cover his third point, about measuring customer satisfaction, in Chapter 6. In this chapter we discuss the rest of the ideas on his list, and then we'll add one further point: the importance of *engaging every person's whole mind in improving your organization.*

Let's look at each of these points in turn and see how they add up to a simple, workable program for freeing the customer champions in any workforce.

■ HIRING: HIRE PEOPLE WHO LIKE PEOPLE

Cawley of MBNA is right on target when he tells us to hire people who "like people," and he tells us how his organization does it: by requiring senior management to interview every candidate. Any manager can serve customers better with outstanding people. And most organizations can bring in much better people than they're hiring now by giving the hiring function the care it deserves.

Two cardinal points underlie Cawley's message, and they apply for every company. In hiring:

- Focus on attitudes
- Recognize just how difficult and vital the job of selecting people is. Invest enough time in the work, especially if you're senior manager.

■ Focus on Attitudes

Nothing matters more than an employee's attitudes. It's easy to fill a job with someone who has appropriate formal qualifications, such as college degrees and years of experience. But too often these "qualified" people don't truly serve your customers.

Zig Ziglar, the motivational speaker, often asks people to write down the characteristics of a good salesperson. They usually write such things as:

- honest,
- diligent,
- hardworking,
- good with people,

- ready to follow through,
- loves customers,
- high integrity,
- knowledgeable about product.

Then Ziglar asks people to list how many of these characteristics are skills that can be taught or absorbed in years of experience, and how many are attitudes, which are much harder to teach. Most of the characteristics are **attitudes**.

Try the same exercise for *any* other position—say, a factory worker, customer-service representative, manager, research scientist—and you'll achieve remarkably similar results. Of course, you can't hire research scientists who lack graduate degrees in the sciences. But you'll pick winners if you remember that graduate degrees are just basic qualifications—the people with the right attitudes become the great producers.

■ Invest Enough Time

The time that managers invest in hiring pays big dividends. This truth was certainly evident in the success of Nordstrom, the Seattle-based department-store chain that achieved more than sixfold growth in the 1980s. Other retailers often accept mediocre people willing to tolerate mediocre wages, but Nordstrom applicants go through as many as three interviews before they're selected. To hire 400 people in its new McLean, Virginia, store, Nordstrom's interviewed 3,000.[2]

In 1989, to be sure, Nordstrom suffered a setback when the State of Washington claimed the company had failed to pay employees for some of the time they spent providing superlative customer service—personally delivering goods to customers' homes, for example. The company was ordered to pay back wages.[3] A mistake, yes, but it shouldn't obscure a vital lesson. Nordstrom showed the retailing industry that finding and motivating excellent people was both possible and a way to superior profitability. And if it can be done in retailing, it can be done in any industry.

■ How to Hire: Make a Profile and Stick to It

At the beginning of the hiring procedure, carefully create a profile of the person you're seeking. Deliberately review the characteristics of people who have succeeded in your organization, and aggressively pursue them in applicants. Include the attitudes and skills the job itself requires and also the attitudes your organization's vision demands. Even a purchasing clerk or a lathe operator should want to delight customers—both the organization's ultimate customers and the "internal customers" the individual serves.

Creating an excellent profile is hard work and may require guidance from outside sources. A study by my own organization revealed, among other things, that candidates who will become outstanding salespeople exhibit two behaviors that are easy to spot in an interview. They show enthusiasm, and they make eye contact.[4]

Once a profile is created, stick to it. Ask questions that reveal the applicant's past **behavior** in situations in which he or she should have shown the attitudes you're seeking. Robert Gilbert is manager of the Rohm & Haas Bayport chemical plant in LaPorte, Texas. His plant is famous for relying on workers' initiative—it has no shift supervisors. Gilbert thus seeks people who will take responsibility and care about others. He asks applicants:

> When was the last time you saw somebody doing something dangerous? What did you do?

When you talk to references, ask if the candidates think about service to people and if they're team players. Ask all the other questions necessary to determine not just whether the candidate is a decent person, but whether he or she really matches your profile.

Also, when checking references be sure to go back three jobs. Most recent employers are concerned about the legal risk of an honest assessment or are eager to have the person they fired get a new job. In either case, the data you get on your candidate will be filtered and often misleading. The further back you go, the more honest former employees tend to be.

■ The Boss Who Doesn't Understand

I've been making speeches on how to do right by customers for years. I've probably given more than three hundred. And after I make a speech, people ask questions.

Unfortunately, about 70 percent of the questions all sound the same. They go something like this:

> What you said up there makes a lot of sense to me. I believe it. But I can't get my senior executive—the manager who's above me in my company—to pay any attention.

I've given a lot of thought to answering that nagging question. I suggest three actions to get through to senior executives who don't understand or don't care about truly serving customers.

1. Have executives spend time with customers

The first is to **get them out talking with customers.** Make them see and hear the people who are buying your products. No manager will flat-out refuse to visit customers. And it can make a big difference.

One manager who heard me speak in Boston was from Millipore, a company that makes filters. She addressed the audience and said:

> You know, I had a real problem getting changes going. And what I finally had to do was to take our CEO and drag him out on these customer visits. We actually went to or met with 100 different customers. That made a believer out of him. I now have an unbelievable support system in that organization that starts from the top. I cannot tell you how powerful this is.

Just encourage managers to listen to customers talk about their problems. Also encourage customers to tell them the good things that a few people in your organization are doing. That has an incredible effect.

Another businessperson made this comment:

> You know, our experience is that this all comes down to two things: passion and persistence. Of the two, passion is probably more important. You've got to light a fire under your executives.

One of the best ways to do that is to send them out to be with customers.

2. Do It with Data

The second approach is: **do it with data.** Collect good, hard statistics that show how much your company is losing because it doesn't take good care of its customers. Then if your managers aren't used to thinking of customers' real needs, that won't matter. They've spent years learning how to read and analyze business statistics and charts. This is, in a way, a business language—their business language. Speak it.

If you tell a group of hardnosed executives, "We've got to empower our people and work in cross-functional groups," you'll scare them; you might as well be talking in Swahili. But if you support everything you say with good, solid data, you can win them over.

3. Benchmarking

The third approach is to **use comparisons and benchmarks.** This, too, is astonishingly powerful, arouses passion. There's not only a corporate desire to perform well, but also a sort of corporate ego. "How's the next person doing? How are those competitors over there doing?" When my organization conducts surveys across numerous organizations, we always go back and validate. We go back to the organizations that participated and we tell them the results.

You can guess what all the managers focus on. It's the charts that say: "This is where you stand; this is where others stand." When we go into organizations, we can't believe how powerful that is. People don't even want to talk about whether the analysis is valid. They just want to look at the comparisons. The comparisons are a way to get them off dead center, a way to get them rolling. Let them go visit a competitor or somebody who is not a competitor but has the kinds of abilities that you'd like your organization to acquire. See Chapter 3 for more information on benchmarking.

Do those three things:

- let them see and hear customers,
- give them hard data, and
- engage in some kind of benchmarking.

Those can turn your managers around.

It's my bet that companies make more hiring mistakes because they fail to focus on attitudes, create clear profiles, and stick to them, than for any other reasons.

■ LET PEOPLE KNOW WHAT TO EXPECT AND WHAT'S EXPECTED OF THEM

New employees and job candidates are looking for a chance to be happy and serve customers. They need to know how *they* fit into your vision. Tell them what to expect:

1. What they should anticipate—good and bad—in their daily work.
2. How you'll measure their performance.
3. How you'll recognize them and reward them for good work.
4. How vital *serving the customer* is in everything the organization does and every rule it makes.

■ Teaching What to Expect Is Good Business

Just telling people, as truthfully as you can, what lies ahead can save much pain for both you and your employees. I was once involved in a study of an insurance company. The firm had a mystery on its hands. Turnover for agents was high—typical in the insurance industry—except at one branch office. There, it was strikingly low. How could that be?

Finally we discovered the reason, and it was elementary. When the manager of this branch talked to people interested in working for him, he pointed out all the good aspects of the job. "You can be independent," he'd say. "You can be an entrepreneur. You get paid for performance. You get to go on all kinds of trips and get all kinds of recognition when you succeed. And you get to help people."

Then the branch manager would pause. "But wait," he'd say. "You'll be working on weekends and evenings. You'll get more than your share of rejection. And, over and over again, you'll get doors slammed in your face."

He told the prospective agent to write down the points—negative and positive—on a card and think about them, then come back if he or she was still interested. The people who decided to become agents had few false expectations. They were ready for both the best and the worst. And they stayed at that company and made the office where they worked unusually productive.

■ Customer-Driven Measurement

Nothing worries people more than how they'll be measured and whether they'll be rewarded for doing a good job. Explain your system for measuring performance to newcomers and to all employees who may not understand it. Ask for suggestions about improving your system, and consider them seriously.

At MBNA, sales managers regularly listen in on employees handling calls. This kind of monitoring can benefit an organization immensely. It enables managers to understand exactly what the customer faces. And it allows them to advise the people handling the calls on how they can serve the customer better.

But managers need to explain a policy like this carefully, and show employees that they're doing it for the customer's benefit. Otherwise, it could seem like Big Brother on the prowl. At MBNA the policy is described in some of the first interviews conducted during hiring, and employees are also given notice in writing.

If you're uncomfortable explaining your system for measuring people, perhaps that's because you haven't developed a good one. Too many companies measure what's easily measurable, rather than what will help the customer. Productivity-oriented monitoring, for instance, can easily discourage, rather than promote, good performance. Evaluating telephone operators primarily by the number of calls they handle encourages rushed (and therefore poor) handling of calls. Many telephone operators hang up the phone in the middle of their sentence:

Have a good d . . .

because they're trying to achieve productivity norms. Similarly, poorly designed piecework systems and productivity-oriented compensation for salespeople often discourage high-quality, customer-satisfying work.

At MBNA, on the other hand, customer-service representatives are told that once they've taken a reasonable minimum of calls they'll get no extra credit on their performance evaluation for taking more. Other outstanding organizations, such as General Electric's Answer Center, do reward people for handling as many calls as possible. But they also survey customers carefully to learn whether the representatives are satisfying them.

■ Rewards: They're Not in It Only for the Money

Rewards—compensation, recognition, and promotion—are among the strongest statements you make to your people about how you value them and their work. If you don't communicate accurately about your reward and recognition system, or if the system itself doesn't convey customer-driven messages, you're missing a powerful opportunity to link people's behavior with your customer-driven vision.

Compensation

When you reward a manager for "making the numbers" in a quarter or a salesperson simply for exceeding a quota, the organization will draw powerful conclusions about what's important to you. Thus you face an exceptional opportunity but also exceptional dangers when you create compensation policies and communicate them. To the extent that you can make people aware that you'll reward them for behavior that truly serves customers, you'll vastly strengthen the power your vision has throughout your firm.

Perhaps all systems of compensation can somewhat misdirect some people; whatever numbers you collect and whatever approaches to measurement you use, a few will always try to beat the system and figure out ways of looking better than their actual performance for the customer would justify. But you *can* create a system that at least moves most people to serve the customer better. Often this system may entail basing incentives on long-term rather than short-term performance, and creating group as well as individual incentives.

The MBNA organization uses group incentives. An index reflects fourteen measures of its customer-service people's performance (such as how often the phone is answered within two rings). It pays a healthy bonus to all employees every quarter, depending on how well those objectives are met.

Often you'll achieve the best performance by creating small work groups and setting team incentives. Worthington Industries, the most profitable company in the American steel industry during the 1980s, works to limit the number of people in any one workplace to 100 or fewer. The president of Don Rodman Ford in Foxboro, Massachusetts, one of the most successful Ford dealers in the world, found that customer satisfaction increased dramatically when he organized his mechanics and service people into small, dedicated teams. Each team is big enough to handle almost any kind of customer problem, small enough so that responsibility is clearly identified. Customers feel secure knowing who's working on their cars.

Many organizations overemphasize compensation. Strangely, giving away money can actually hurt key elements of performance for the customer. David Luther, senior vice president for quality at Corning, says this about the suggestion program his company had in its factories until the 1980s:

> In our old program we'd give people 10 percent of the value of the suggestion. Now suppose you're an employee. You come in and say: "Here's an idea for a typewriter, and it's going to save about a million bucks." I look into the idea and maybe it will save $1,000. You're looking for a check for $100,000. I'm the supervisor and I'm talking about $100. That kind of thing can cause a lot of mistrust to develop pretty quickly.

If you have a choice between creating destructive incentives or no short-term financial incentives at all, you're better off without financial incentives. People will try to do the right thing as long as you don't actively discourage it.

Recognition

A powerful management opportunity lies in creating ways of recognizing good performance without necessarily paying a lot of money. Today Corning has a new suggestion program, rewarding people not with cash but with plaques, company jackets, special dinners, and local publicity. It generates about *forty times* as many suggestions as the old system did.

One survey of Metropolitan Life employees showed that, given a choice, they actually preferred management recognition for good service to cash awards. Most employees, unfortunately, don't know their companies appreciate them.

Promotion

If rewards are the strongest statements you make to your people, promotions make the strongest statements of all rewards.

I often say to executives:

Think of the last five people you've promoted. They represent what you value. What message have you delivered to your entire company with these promotions?

Often, managers promote people because:

- they've been around for a long time and "paid their dues,"
- they've never offended anyone and the promotion is therefore "safe,"
- they're friends with the right people,
- they're highly competent in their technical specialty (which may have little relevance to performance in the next job), or

- they've played office politics effectively, so that they appear brilliant to the boss even though others in the organization don't respect them.

Employees see these promotions and conclude that these are the qualities the company values. The promotions have said nothing about the importance of serving the customer. Find some other way to acknowledge the person who's been around a long time and kept out of trouble. Let your people know you'll give promotions to people who **embody your company's vision.**

■ Show People You Expect Them to Serve

Telling people what to expect day to day and how they'll be monitored, rewarded, recognized, and promoted should all point people in one direction: toward an understanding that *creating happy customers* is the heart of your organization. When you let people know this principle and make it real in your own behavior, you encourage excellent performance and low turnover.

The Forum Corporation's Customer Focus Research study showed that the factors correlating best with turnover of service personnel were their responses to questions like:

- Does the company mean it when it says it's committed to customer satisfaction?
- Does the company ask me for my ideas on how things could be improved?
- Are people around here trained in multiple jobs?

In short, *knowing that they are expected to use their abilities to serve the customer* makes employees feel they are part of the action and encourages them to stay with the firm.[5]

■ EDUCATE PEOPLE FROM THE TIME THEY'RE HIRED

Educate people well and continuously. Then you'll get the most they can give.

Education is one of the greatest strengths of the Japanese corporation. A study by Massachusetts Institute of Technology Professor John Drescheck revealed that Japanese auto workers got two and a half times more formal training than their United States counterparts. Including informal training such as help from a senior worker, the Japanese worker got three times as much as the American.

Today Western companies can't afford to allow their past noninvestment to continue. Continuous formal education is vital for three reasons:

- It provides an opportunity to communicate and reemphasize the company's vision.
- It provides concrete skills that employees need.
- It shows people alternatives to old ways—alternatives from which they can create their own new ways of doing things that serve the customer better. Thus, it opens the organization to improvement.

■ What Is Good Training?

Good training is essential in transforming a company so that it will serve the customer. Good training enables people to engage in and sustain the right kind of behavior on the job—quickly. A person needs training most urgently when he or she isn't delivering the behavior that a customer-keeping organization needs. "Customer-service agents" may have the word "service" in their job title, but they often think of their jobs as clerical and treat real live customers as an annoyance. Remember John the hotel clerk? You'll help your people best if you have clear objectives, defined in specific behavioral directions:

- First, understand how people are behaving *before* the training.
- Then, define how you want them to behave when training is completed.

Teach knowledge, teach real skills, and teach your corporate vision. Don't lecture people about attitudes. People don't respond to courses that just tell them they should smile at customers. But if your organization shows it really cares about customers in other ways, and if your training

---

■ **What Do Your People Need to Learn?**

To diagnose employees' performance—to figure out why the employee isn't doing what should be done—you might try a simple method devised by Robert Mager, a training consultant. It consists of asking five questions.

1. Does the person know what he or she is supposed to do—his or her basic mission in your organization? The hotel clerk who didn't answer my question didn't know that his basic mission was to make customers happy.

2. Does the person know the tasks he or she should do to accomplish that basic mission? The job of any member of an organization demands taking specific steps. But many employees don't know exactly what those steps are.

3. Does the person have the skills to perform those tasks? Perhaps the person doesn't know how to run a facsimile machine, and so leaves customers fuming through a trial-and-error exercise every time someone wants to send a fax. Or perhaps the person lacks time-management skills, and wastes time simply waiting on duty rather than preparing for predictable customer needs.

4. Does the person have sufficient resources to do the job? Often companies lack systems to deliver even such basic resources as computer printer ribbons to front-line employees.

5. Could the person do the job if his or her life depended on it?

If the answer to questions 1, 2, and/or 3 is No, you have a training problem. If the answer to question 4 is No, you have a resource problem. If the answer to question 5 is Yes, you have a motivational problem.

---

programs teach a vision of service together with real skills that will help people attain that vision, you'll change attitudes in the bargain.

Good training today is better than good training even ten years ago. We know much more about how to design a program than we used to. Most people will try to teach by *telling* people how to do something, for instance. But research has clearly demonstrated over the past two decades that the most efficient way of teaching people is to get them actively involved with the subject. Before middle managers come to a session on quality, have them tour your factory. If you're teaching customer service, have employees interview customers.

When evaluating trainers, either company staff members or an outside firm, ask three questions:

- Are these people asking me about my needs and objectives, or do they seem more interested in presenting pat solutions to what *they believe* my problems are?
- Can they explain how they will transform people's behavior?
- What results should be measurable in job performance? Should employee turnover decline? If so, by how much?

Then, after the first round of training is complete, ask: Did people's behavior actually change as promised? For a quick test, do some role playing with the people who were trained. To see if your managers learned how to listen better to the people who work under them, see if they listen better to you.

The Walt Disney Company is a leader in providing good training, and it has reaped the benefits on the bottom line. Walt Disney himself started the Disney University before Disneyland opened in Anaheim, California, in 1954. He realized that the traditional amusement park—a collection of thrill rides serviced by often-obnoxious fast-buck artists—wasn't what the customer deserved. He trained people to be part of a live, three-dimensional, constantly continuing show—a clean, engaging family experience.

Everyone from park sweeper on up takes at least four days of courses that include:

- "traditions." What is the Disney vision? What achievements through-out the company's history have embodied that vision?

- basic information. If a guest asks a sweeper where where a ride is, the sweeper will know.

- how to act. What do you do if a guest asks you a question and you don't know the answer?

The result is a group of employees (referred to at Disney parks as cast members) who differentiate Disney strongly from ordinary amusement centers.

Disney added substantial value with his training. Theme parks now produce three times as much profit as all the Disney Company's movies and cartoons put together. Even Frank Wells, president and chief operating officer of the Disney company, went through Disney University's orientation program—after joining the company with a multi-million-dollar contract.

■ CREATE AN ENVIRONMENT THAT SUPPORTS THE PURSUIT OF CUSTOMER SATISFACTION

The most admired, most profitable companies share a common denominator: happy people. Successful business leaders provide the resources their people need to serve customers well.

Take a tour of your office or your plant. The way in which you treat the work environment is the way in which you're encouraging the people who work there to treat the customer. John Sewell, of Sewell Village Cadillac, the prize-winning dealer in Dallas, found that a seemingly innocuous decision he made turned out to be highly significant. He decided to paint the walls of his repair shop white and add skylights. He did it to create a clean and bright place where customers could leave their cars to be repaired. But it dramatically affected the mechanics who worked there. They appreciated the improved environment, and as a result, he says, they did better work.

Just as the first step in serving customers is to find out what they really want, a major step in giving your employees the environment they need is to learn what's important to them. It's especially helpful to understand employees' desires when you seek to reward them for doing good work. Kodak has begun to ask employees: "How can I reinforce you when you're doing a good job?" One mailing-room group said it wanted quality specialist George Vorhauser to buy them coffee and donuts if they completed a particularly arduous mailing without an error. They did, and he did.

"Different people get turned on by different things," Vorhauser writes. "Back in 1981, if the boss brought in donuts, it just would have indicated that the boss wanted to do donuts." Even employees who enjoyed the donuts wouldn't have been as favorably affected as they are now that managers ask them in advance what they want and make clear that the reward is a direct result of their performance.

■ TREAT PEOPLE LIKE CUSTOMERS

If you are a manager, then every individual in your work group is one of your internal customers. If you are a top manager, then everyone in the company is one of your internal customers. If you are making a decision, preparing a policy statement, or circulating a questionnaire, your work has to be acted on by the others in the group before it can benefit customers, stockholders, or anyone else.

Therefore, treat the people who work for you with the respect you'd give any customers.

I'm not quite ready to join MBNA's Cawley in advocating that we completely cease using the word "employees." But I fully agree that it's an unfortunate word. It overemphasizes that a person works *for* someone else, and underemphasizes the reality that we all work *together*. Some companies call their people "members," as at Walt Disney's theme parks, where they are cast members. Other companies, such as the apparel chain The Limited, call their people "associates." Unfortunately, there's no agreement on a word to replace "employees." And sometimes organizations need a word that's more specific than "people."

Whatever words we use, the point is to respect the people who work with us. That leads to happy employees, excellent work for customers, and superior profitability.

ServiceMaster Company, the fast-growing maintenance and food-service firm, has become the most consistently profitable company in the entire United States, according to *Fortune*. One reason: ServiceMaster treats an uneducated workforce with exceptional respect. Many ServiceMaster employees can't read or write. But its chief executive has said: "Before asking someone to do something, you have to help them to *be* something."[6]

▪ Problem Employees

Treating employees like customers doesn't mean ignoring their failings. Problem employees can make customers miserable, and you can't let that happen.

But a problem employee deserves to be treated as you would treat a problem customer. Good businesspeople don't walk away when a customer demands unreasonable services or pays bills too late. They work on the relationship to make it worthwhile.

Work, then, with employees who cause problems. Help them become part of the company vision.

Consider the problems of middle managers, often the biggest obstacle to a customer-driven transformation of any company. John Bray, CEO of Forum Europe, calls the managers two or three levels below the top of an organization "the thermal layer." This group has created a comfortable life for its members by doing things in the old way, and so they fight against any transformation. Top managers feel their efforts to improve the organization disappear into the unresponsive thermal layer almost like pebbles thrown from a bridge.

■ The Cost of Turnover

Companies take better care of their people when they realize how much it costs to replace every person who leaves. With the number of youngsters entering the workforce declining sharply, those costs can only grow. You'll have a head start if you understand them.

List the costs your organization suffers from turnover, and calculate the actual cost for each item. You'll be astonished at the results. Here's a start:

1. loss of management time invested in training the departed employee;

2. pay given the former employee while he or she was learning the job;

3. cost of lost opportunities and customer dissatisfaction while the job is vacant;

4. recruiting costs and fees to find a replacement;

5. management time spent interviewing candidates for the vacated job;

6. cost of training the new employee;

7. salary paid to the new employee before he or she really knows the job;

8. cost of extra time the new person's supervisor must invest in him or her;

9. cost of mistakes the person makes when first on the job;

10. disruption suffered by the office during replacement;

11. loss of knowledge of the business that only the departed person possessed.

You can probably add half a dozen costs that pertain to your organization.

A few organizations claim that turnover isn't a problem. Fast-food companies sometimes argue that the jobs in their organization are so simple that they can easily train replacements, and that people who remain long in those jobs lose enthusiasm.

Each organization is different, but I wonder whether such companies really understand what's best for them. If people lose enthusiasm, why aren't their managers putting their talents to use in a way that would retain it? Why aren't they asked for help in improving the organization? Why can't experienced part-time employees be promoted to part-time assistant managers? I've seen lots of fast-food restaurants where service to customers is disorganized, and they could use some extra assistant managers.

Whatever business you're in, make sure you fully understand the costs of turnover. Your employees, your customers, and your bottom line will all benefit.

Many companies, though, make this confrontation more adversarial than it has to be. They create "employee participation" schemes without inviting participation by the middle managers who will have to make them work. Middle managers' natural skepticism rises even higher.

The past decade has shown that smart companies can lead middle managers even if they do resist change. These companies involve middle managers in the effort to change. Middle managers are represented on the team that plans the whole initiative. When these organizations have gained the commitment of top managers and the rank-and-file employees—who are often much easier to persuade that change has advantages—the real achievements and the enthusiasm of the top and bottom can gradually win over the thermal layer. However, publicizing those gains achieved by middle management themselves will influence them the most.

■ The Power of Well-Treated Employees

The last time I was at the Marriott Camelback Resort in Scottsdale, Arizona, an assistant manager volunteered to pick me up at my room and drive me to the tennis court on a golf cart. There was no reason except that we had gotten into a conversation in the lobby. I asked him, "Why do you go out of your way for me?" His answer floored me:

Well, I guess it's because the company goes out of its way for us.

Here was an individual taking time out of his day just to focus on me. This attention made me feel special. And it was there because someone made him feel special.

■ "Think of Yourself as a Customer"

Think like your customer, and that attitude will be transmitted to the people who work with you.

Even though it faces one of the most difficult customer-service tasks, MBNA is succeeding. It has nearly 8 million customers scattered all over the United States, and its people never see them.

Why then do they serve the customer well? The reason is that each element of MBNA's approach to managing people contributes to a community that thinks like its customers and champions their needs.

■ A FURTHER POINT: ENGAGE YOUR PEOPLE'S MINDS

One crucial aspect of people management today might be overlooked in Charles Cawley's MBNA program: engaging people's talents not just in doing their regular jobs for the customer, but also in constantly improving the company.

Companies have long tended to hire people for their brawn. Today, we need every person's brain as well. And there's good news. People are happy to put their brains to use for their companies and their customers.

When we ask someone to do a job, all we need to do is say:

Here's the best way we know of doing this. We'd sure like to know what you think.

Take Rohm & Haas Bayport, the chemical plant in Texas. All sixty-seven employees have substantial roles in managing the plant, even evaluating one another and interviewing job applicants. "We're trying to transfer ownership for decision making to the people who are going to get the work done," says the plant manager, Robert Gilbert. And Rohm & Haas Bayport has achieved such outstanding quality that it received the highest rating ever given in one customer's review.

At Baxter Health Care Corporation, a truck driver in the Atlanta region formed a corrective-action team. Twenty percent fewer trucks were needed after the team simply rerouted and rescheduled pickups.

At Bank One in Columbus, Ohio, a team with people from several departments cut check-clearing time almost in half.

And at Fuji Electric of Tokyo, the *average* employee submits 277 suggestions a year. About 98 percent are put into service, and 80 to 90 percent of those are implemented within a week by the individual's supervisor without higher authorization.

Fuji has a policy of acting not only on suggestions that are clearly good for the company, but also on those the supervisor believes are neutral in their effect. Most people in the company have a chart at their workplace showing the number of suggestions made and implemented.

All we have to do is ask. We don't even have to pay our people more money for their ideas. People constantly surprise successful managers by showing they can and will do far more than any old-fashioned boss would have imagined.

If you give everyone a chance to use his or her mind, you'll not only serve the customer better, you'll also discover the hidden skills that make people promotable. That's a big bonus in a world of tightening labor markets.

■ People Want to Become Customer Champions

Even some of the worst organizations made a remarkable discovery once new management arrived. People within them had wanted to serve customers and had the ability to do it well all along. They simply needed basic support from customer-driven leadership: an unswerving commitment to customer satisfaction, constant communication of that objective, careful measurement, intelligent hiring practices, continuous education, a well-managed environment, respect, and the chance to use their minds to improve the workplace.

Says David Gunn, who as president of the New York City Transit Authority turned an ugly, undependable system into something riders could rely on,

> We had a workforce that was absolutely willing to do the right thing. The problem was management.[7]

Today, millions of employees want to be customer champions. Liberate them with vision and opportunity, and they'll do the rest.

ACTION POINTS

- Devote enough management time to hiring, and emphasize *attitudes that are consistent with your vision* more than than formal qualifications. Make a rigorous profile of each person you seek, and stick to it.
- Tell everyone in your organization how they'll be monitored, and adjust your monitoring system so that you're proud to explain it to your people.
- Review your reward systems, the systems that make your "most-listened-to" statements. Ensure that they communicate that you're serious about your customer-keeping vision.
- Educate people continually. Introduce training programs that create measurable, predictable, constructive changes in your people's behavior. And if the programs don't deliver as promised, find programs that do.
- Create a work environment that shows you care about your employees in the same manner as you would have them care about customers.
- Treat problem employees as you would treat problem customers—as business partners with whom you want to create a profitable relationship.
- Engage your people's minds in constantly improving your organization.

■ RESOURCES

Peter Block, *The Empowered Manager* (San Francisco: Jossey-Bass, 1987).
John Gardner, *Self Renewal* (New York; Norton, 1983 (1964)).
Rosabeth Moss Kanter, *When Giants Learn to Dance* (New York: Simon and Schuster, 1990).
Harry Levinson, *The Exceptional Executive* (Cambridge, Mass.: Harvard University Press, 1968).
Tom Peters and Nancy Austin, *A Passion for Excellence,* (New York: Random House, 1985).
Peter B. Vaill, *Managing as a Performing Art* (San Francisco: Jossey-Bass, 1989).
Robert H. Waterman, *The Renewal Factor* (New York: Bantam, 1988).

■ NOTES

1. National Quality Forum meeting, New York, October 3, 1989.
2. Rebello, Kathy, "Seattle Chain Opens Today in Virginia," *USA Today,* March 4, 1988.
3. Susan C. Faludi, "At Nordstrom Stores, Service Comes First—But at a Big Price," *Wall Street Journal,* February 20, 1990, p. 1.
4. William M. DeMarco and Michael D. Maginn, *Sales Competency Research Report,* Forum Corporation, p. 13.
5. Forum Corporation, *Customer Focus Research Study* (1988), p. 35.
6. Heskett, J. L., "Lessons in the Service Sector," *Harvard Business Review* (March/April 1987), p. 118.
7. Daniel Akst, "Where Did All the Graffiti Go?" *Forbes* (May 28, 1990), p. 334.

Simplify, and goods will flow like water.
—Richard Schonberger,
Japanese Manufacturing
Techniques[1]

5

Smash the Barriers to Customer-Winning Performance

There it is in figure 1, newly discovered and hard to see, and often fatal to companies and other organizations.

It's called "the iceberg of ignorance." It surfaced when a research group in Japan under the guidance of quality expert Sidney Yoshida asked a cross-section of people in a large factory to list all significant problems known to them.

Surprise: Only 4 percent of the problems listed were known to top managers. Most of the causes of defects, excess costs, delays, and other problems for customers were hidden, like the vast mass of an iceberg.

Figure 1 shows who really knew what was going on. It reveals that virtually every problem known to top management was also known to at least a few of the rank-and-file employees. What's more, foremen and other employees knew of hundreds of problems of which higher-level people were blissfully—and dangerously—unaware.[2]

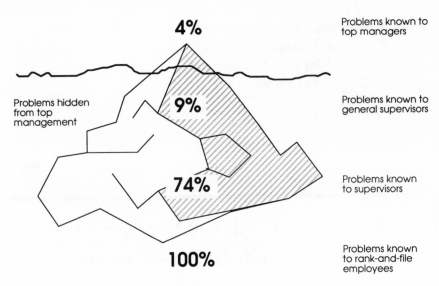

4% — Problems known to top managers

9% — Problems known to general supervisors

Problems hidden from top management

74% — Problems known to supervisors

100% — Problems known to rank-and-file employees

Figure 1   The Iceberg of Ignorance

The Iceberg of Ignorance study demonstrates why only a few companies consistently fulfill their promises. Your organization always faces hundreds more problems than you as a manager can see, much less solve.

And yet a relative handful of organizations and workgroups continually amaze customers with their reliable performance and their ability to deliver what they promise. Rubbermaid, Toyota, Disney, and others achieve this performance unfailingly.

In this chapter we'll show how they do so—and how you can do it, too. Reliability and customer-winning performance, we'll find, isn't necessarily achieved just because everyone is "doing their best." Usually one basically simple strategy appears:

Eliminate the barriers that prevent your people from serving customers efficiently and predictably.

Eliminate barriers to serving the customer and you get both customer satisfaction *and* lower costs.

You would be hard pressed to find a company with higher quality than Toyota. Look at the reason. Long before Toyota cars were even available outside Japan, Toyota president Kiichiro Toyoda (1894–1952) had his employees hunting down and eliminating barriers that kept their system from serving the customer. Today Toyota uses highly sophisticated, technologically advanced methods to eliminate barriers. For one thing, Toyota leads the world in research that is creating new materials for lighter, faster, and more responsive cars.

But be assured that Toyota does not pursue sophistication for its own sake. Its pioneering Toyota Production System typifies its efforts more than any high-technology research. The system, developed in the 1950s, for the first time allowed customers' needs—rather than rigid production schedules—to drive every part of a manufacturing system. And it achieved that goal simply by looking for wasted effort and finding ways to eliminate it.

Toyota engineers asked: Why take a part from a supplier and carry it to a storeroom, only to take it out of the storeroom again when it's needed on the production line? Why not set up the system so that the parts are carried directly from the supplier's truck to the assembly line and arrive there just before they're needed? Wasted effort, a barrier to efficient production, was eliminated. So too was storage time during which parts could deteriorate, and warehouse space used for the storage. Everyone at Toyota got used to making decisions based on facts, not hunches.

For decades the Toyota system has methodically destroyed even minute obstacles to serving the customer. Thus Toyota products are perennial favorites in customer-satisfaction reports. Barrier smashing also has yielded efficiency. Toyota's customer-driven system produces twice as many sales per employee as Nissan, and two-and-a-half times as many per employee as Chrysler Corporation.[3]

This chapter will show how your organization can continually eliminate barriers to customer satisfaction. We'll discuss several elements in the procedure:

1. First, we'll look at the causes of the problem. Why do most companies fail to fulfill their promise?

2. Then we'll show what happens when companies adopt very simple techniques to eliminate barriers: such techniques as asking employees and suppliers for ideas, or asking customers to tell you their problems and then assigning someone to find and eliminate the causes.

3. We'll explain a simple six-step problem-solving procedure that can work on almost any problem.

4. We'll look at advanced yet fundamentally simple systems that can make customers' needs drive everything within a highly complex system such as a factory.

5. We'll discuss what's special about service companies—how the problems they confront resemble those of manufacturers, how they differ, and how service companies can deliver true dependability.

6. And finally, we'll look at crucial methods of controlling a company as a whole so that it will constantly eliminate barriers in all these ways, and avoid creating new ones.

■ LOST IN THE SWAMP

The fundamental reason keeping goods, services, and innovations from "flowing like water" is that people are bogged down, spending much of their time and energy dealing with problems they believe are unavoidable. High-performance companies, focused on efficiently serving the customer, eliminate the problems that others believe are inevitable.

Goods in the factories of the best-led manufacturers flow toward the customer like a free-running stream. But the factories of ordinary companies—and the service practices of most companies, as well—are more like swamps. As organizations grow, barriers to serving their customers grow like weeds. Most barriers—confusing management structures, manufacturing methods that allow machines to produce defects, inability to manufacture in small production runs—are completely unnecessary. They produce waste without compensating benefit. Directives, slogans, and ordinary cost-reduction campaigns can't eliminate them. They persist, and they create major problems.

Because of barriers, "swamp" factories take many times longer than "stream" plants to effectively adopt new technology. Their confusing systems make it hard for anyone to focus on the customer.

And sadly, most service providers are even more bogged down by barriers than factories. Although many applied, not until the third year of the Malcolm Baldrige National Quality Award, were service companies among the winners. (Federal Express and Wallace Company, a distributor of oil field equipment, finally won in 1990.)

In both service and manufacturing companies, policies and procedures are too often designed for the convenience of an internal department rather than for the customer's benefit. This arrangement is somewhat analogous to that of the nurse who wakes a patient to administer a pill—not because it's time to take it but because the nurse is going off shift. Neither the companies nor the nurse are driven by a customer-focused vision.

■ ASK FOR HELP, AND THEN IMPROVE PROCESSES

The most elementary method for breaking down barriers is pretty simple:

> Ask employees, suppliers, and distributors to tell you the barriers
> they see preventing your organization from serving its customers.
> Ask for suggestions on how to get rid of them. Then eliminate
> those barriers.

As we've seen with the dreaded Iceberg of Ignorance, people closest to the action—on the factory floor, dealing directly with the customer, or even carrying out accounting transactions in the finance department—often have terrific ideas in areas traditionally held by highly touted experts. These people spend their days working with the nuts and bolts of a business, and they inevitably do a lot of thinking about them.

Consider one project in which these successes were particularly well documented: the development of the Ford Taurus. Taurus engineers had planned to stamp out the side panels of the Taurus in six pieces, then have assemblers weld them together. But a machine operator from Atlanta said it could be done in fewer pieces. He was right. Six pieces were reduced to two, and the savings in welding costs alone were immense. Other Taurus assembly-line people asked questions like these:

- Why do we have to use three different welding guns on the body assembly line, all suspended from the ceiling and getting in our way? Engineers designed a single gun, saving assembly time, equipment costs, and reducing the potential for mistakes.
- Isn't that bolt on the hood assembly unnecessary? Engineers realized that it was, and eliminated it. They saved money on bolts, reduced assembly time, and left one less part that could cause problems for the customer.
- Why do we have to use different sizes of screws to fasten interior plastic mouldings? Engineers agreed on a single size of screw, saving both procurement costs and production time.

Ford made more than 700 changes like these based on rank-and-file employee suggestions in the Taurus program alone. One Ford executive estimated that a typical recommendation saved $300,000 to $700,000. Nearly all of them also improved quality.[4]

Suppliers working on the Taurus project made these contributions:

- recommended and helped to design a retractable picnic table as part of the station-wagon tailgate.
- pointed out a method for laying carpeting to give a more uniform, luxurious appearance.
- suggested dual sun visors on the driver's side of the car, one in the normal position in front of the driver and the other above the door. The customer no longer has to constantly fidget with a single visor depending on the sun's position.[5]

Ford reorganized the entire design process to break down barriers among the market research, design, engineering, and manufacturing groups. Was it worth the effort? The Taurus was the most successful American car of the 1980s.

But can we make that kind of performance routine without investing so much time that other parts of the business suffer? The evidence suggests it's possible. Many Japanese companies get hundreds of suggestions each year from the average employee.

Not many companies, though, are eliminating barriers as efficiently as they could. Every manager can greatly increase the success of his or her organization by seeking knowledge from the people who have a massive store of it—the people in their department or group.

- **Ask** each person for suggestions.
- **Specifically** seek suggestions that address the biggest problems your research shows customers are experiencing.
- **Monitor** how many suggestions are coming from each person. (At Fuji Electric, a chart over each person's workplace shows the number of suggestions made.)
- **Promptly implement** not just suggestions that seem likely to improve the business, but also those which seem at least neutral in effect. (At the very least, the person who made the suggestion will feel he or she has been recognized as a contributor.)
- **Recognize** the people who make suggestions.

Whatever you do, don't make the same mistake as one company I visited. They began a suggestion program and ended with such a backlog of ideas that, I was told, it took up to six months to get back to the suggestor. You can imagine the disappointment, resentment, and mistrust this bottleneck engendered.

If you're not getting suggestions that help you serve your customers, communication between you and your associates has broken down. For your customers' sake, it's worthwhile to get it working again.

■ Enlist the Customer's Help

You'll truly fulfill your company's potential only if you include customers in your decisions about barriers. One reason Boeing has achieved dominance in the commercial airplane business is that its worldwide field representatives always relay information back directly to the engineers. They see this as primary to their jobs. The engineering department responds in kind by keeping a close track of the information and acting on it.

Acting on customer feedback isn't always easy: Customers often don't understand your business as well as employees or suppliers. Better than anyone else, though, they know what's not happening for them that should be. That's the place to start. They know what's wrong but it's your job to find the barrier to get rid of it.

■ THE BASIC PROCESS OF BARRIER SMASHING

Responding to customers' interests means learning how to communicate problems to the right people, set priorities, find, eliminate, and prevent the barriers causing the problems, and then improve still further. Everyone in your organization must understand this procedure.

It's similar in both manufacturing and services. According to Susan Robinson, director of training at Xerox of Canada, that company has identified "basic business and administration processes" that together make up the whole of its business. In the distribution part of the business, the processes are "the selling process," "the invoicing process," and "the servicing process." Xerox of Canada is analyzing and improving those processes in essentially the same way as Xerox's manufacturing people improve manufacturing procedures.

A basic approach to improvement involves six steps:

1. Collect information, especially from the final customers of your product or service and the internal customers of the business processes that may be causing the problems.

The Customer-Driven Company

Use the techniques discussed in Chapter 2 and also, where necessary, the survey techniques we'll discuss in Chapter 6.

2. Convert customer information into measures.

Two kinds of measures are useful:

- attribute measures, which strictly classify data into two categories: okay and not okay. An order was either taken accurately or it was not. The delivery was either on time or it wasn't.
- variable measures, which report data on a scale. Variable measures are less likely to depend on subjective judgments than attribute measures. (If a delivery is within twenty-five seconds of the scheduled time, does that make it late?) To create a concrete variable measure, therefore, choose a unit of measure that's appropriate to the customer's interest. To meet the customer's expectation, "No unreasonable wait to check in," we'd measure waiting time in "minutes in line." To meet the expectation, "on-time delivery," you might measure delivery time in minutes, hours, or days before or after the promised delivery date.

Further examples of measures appear in Chapter 6 (page 149).

3. Analyze the current process.

Set a preliminary goal based on your customers' desires. Then look closely at the current method. An extensive selection of methods for analyzing processes and solving problems appears in the toolkit at the back of this book, starting on page 229. Process mapping, page 272, is particularly useful at this stage.

4. Design an improved process.

Review your preliminary target for improvement, using your new understanding of the process. Is it ambitious enough? Will it truly improve something that is a high priority for your customer?

Next, design and map an improved process. Ways of improving a process that has been causing problems include:

- **Creating more consistent input.** If you're improving the method for taking orders, can you create a new form to ensure that all order data are taken in the same way? Can you reduce the number of suppliers—one source for a service rather than three or four?

- **Reducing handoffs.** Look at the process and count the number of times one person or group must turn over work to another. Each is an opportunity for misunderstandings, delays, and errors. Can a process be designed with fewer handoffs?

- **Combining steps.** Determine where two or more steps can become one.

- **Performing steps in parallel rather than serial order.** Save time by having different people do different tasks simultaneously rather than carrying out each step in a sequence.

- **Adding value.** Eliminate waste. For example, reduce the distance to transport work in process from one workstation to another. Or increase the benefits the process provides.

- **Correcting existing measures or adding new measures to the process.** Consider whether the process is out of control because it lacks measures or includes measures that are inadequate.

- **Using technology.** Can part of the process be automated? Often, this step means making better use of existing resources, such as personal computers, rather than acquiring new ones.

- **Involving key people early.** Bring people who will be critical to the process into discussions early.

When you finally have a process you think is better, perform an acid test by asking: "In what ways does this proposed process actually add value for customers?"

5. Establish standards.

Devise process standards so that essential results and accurate measurements are clearly defined. Define both the *results* expected and the *performance that must be achieved at key steps in the process* to achieve those results consistently. "Delivering parts to the customer on time, every time" would be a customer-focused results standard. "Delivering customers' requisitions to the warehouse within one hour of receipt from the customer" would be a process performance standard.

To assess any standard, use the four criteria represented by the acronym MARC:

- **Measurable:** A process standard must be measurable to be of assistance in targeting improvement, and the more specifically you define the standard, the more measurable it will be. Thus, a standard should be measurable by quality, quantity, and timeliness.
- **Achievable:** A standard must be reasonable and attainable, and should be neither so easy that it requires no significant effort, nor so difficult that it requires superhuman effort. In setting standards, you must take into account whether the people responsible for the process have the input and resources they require to meet those standards. For example, a policy stating, "All customer inquiries should be answered within twenty-four hours, not the current forty-eight hours," should be set only when the systems that will allow this change to occur—including access to the necessary resources—are in place.
- **Relevant** to customers (internal and external): The standard must reflect customers' expectations or be designed to create a benefit for the customer.
- **Controllable:** Accomplishing the standard must be within the control of a specific work unit, and the procedure for accomplishing the desired result must be set up.

6. Manage performance.

Review the way in which the people involved in the process are managed and rewarded. In many organizations, management has traditionally stressed meeting quotas, minimizing direct labor, rewarding individual rather than group achievement, and achieving short-term financial goals. Usually this kind of management discourages customer-focused processes.

Evaluate new ways of measuring people, using measures such as:

- quality of output
- team participation
- training hours of employees
- elimination of waste

Stress teamwork over individual heroics. Try to create rewards and recognition that recognize long-term contribution to a team.

■ MAKING COMPLEX SYSTEMS BARRIER-FREE

Many organizations are so beset with barriers that it seems they'll never open the flow so that they can focus on the customer. The more complex an organization, the more the customer can become a vague, far-off entity and the traditions of the people in the organization become the driving force in people's behavior.

And yet even highly complex organizations can be made nearly barrier-free. The Toyota Production System is a good example.

Toyota's system causes everyone in an extremely complicated procedure to perform without waste in a way that will serve the customer. Although Toyota manufactures an advanced product with many customer-chosen options, it now serves as a model for many companies all over the world. It's based on ruthlessly eliminating the barriers to customer-driven action that a complex procedure usually produces.

In complex production or service systems, the number of processes that must be executed is enormous, and "setting up" for each process takes time. Therefore, people try to minimize the time spent setting up by doing jobs in large batches. Production workers make parts in large batches, secretaries order three months' coffee at a time, and loan processors call credit bureaus to request credit reports only when they have a substantial group of applications to process.

But a batch-driven system inevitably creates a morass of barriers to serving the customer. Grouping jobs into large batches slows the entire organization, prevents it from responding quickly to customers' needs. In manufacturing, making parts in large batches means that the results of one process must be stored in inventory until the next process is ready to accept and work on them. In a service function such as processing a loan, operating in batches means that no individual customer can be satisfied until an entire batch of applications can be handled.

When someone makes a mistake, it may not be caught until the entire batch of parts, loan applications, and so on reaches the next stage in the sequence, perhaps days or weeks later. By then, the error is likely to have been repeated many times and the cause may be difficult to determine.

Meanwhile, the large batches of work have to be stored somewhere until the system is ready to carry out the next procedure on them, and storage space costs money. Moreover, keeping track of what's going on is complicated, and extra people must be hired to do it.

One way to improve a complex system is simply to break it into numerous simpler systems. Many companies have vastly improved large workplaces by creating teams, each responsible for completing clearly defined jobs in their entirety.

At Jim Rothmann Chevrolet/Cadillac/Geo in Melbourne, Florida, the service department was reorganized into six teams, each with a leader who could make policy decisions and manage the team budget. A bonus system was established that could give top technicians as much as an additional $100 a week. Those simple steps made a big difference. Employee turnover in the eighty-nine person department went from 70 percent to 8 percent, the dealer's factory-rated customer-service index rose, and profits increased substantially.[6]

Even when the system can't be broken into smaller units, the barriers to serving the customer can still be smashed. Taiichi Ohno, a persistent

Toyota engineer, concluded shortly after World War II that there must be a way to break down the barriers in an automobile factory. He dreamed of a production system in which no one would produce anything until the next process in line—we now call it the "internal customer"—asked for it. A key to making this change, he realized, was sharply reducing the setup time for all equipment in Toyota's plants.[7]

Unnecessary set-up time is one of the worst barriers to serving the customer. Today Toyota and other successful automakers have presses that once took twelve hours to set up; they now take twenty minutes or less. Other machinery that once required thirty minutes now takes perhaps sixty seconds.

Changes to reduce setup time rarely require substantial expenditure. In fact, spending money can guarantee failure. The Buick-Oldsmobile-Cadillac (BOC) division of General Motors reduced set-up time on its giant metal-stamping machines from twelve *hours* to eighteen *minutes* just by soliciting ideas from its factories all over the country. At GM's Chevrolet-Pontiac Canada group, on the other hand, executives decided to reduce setup time by installing billions of dollars worth of new high-tech presses. It wasn't one of the division's most glorious moments. Glitches in the new machinery caused long delays that the low-tech, low-cost solution avoided.[8]

Reduced setup time is the foundation of the Toyota system, often called "just-in-time" in the West. By eliminating the barriers of setup time, the system allows each employee to be driven by the "pull" of customers' demands. It rids the factory of a bewildering maze of batch-production rules that have nothing to do with serving the customer.

Such a customer-driven system eliminates so many barriers to serving the customer that it has tremendous benefits:

- It eliminates the waste that batch-oriented systems produce, allowing lower prices, higher profits, and improved responsiveness to customers' demand. When changes in customers' demand (or even a breakdown in one part of the plant) cause production plans to change under a batch system, no one can quickly adjust schedules for the rest of the plant. The plant thus winds up manufacturing great quantities of unneeded parts. Millions of parts just sit in storerooms, and no one can ensure they'll even be used up before the next model change. But a factory in the Toyota style can function with next to no inventory beyond the parts currently being worked on. Each time a station uses up a box of, say, twenty-four headlight assemblies, it sends a control card or *kamban* back to the group that made the part. That group begins the production process for an additional box. No unneeded parts are produced.

- A plant requires a smaller nonproduction staff when each process is driven by its internal customers. It doesn't need the typical bureaucracy to manage the ordering of batch parts.

- The next process in line within a company—that is, the "internal customer"—can discover and communicate problems in quality soon after an error is made. The worker who is supposed to install a part discovers it's bad, stops the production line, and notifies the person who produced it. The producers find a way to correct and prevent future defects. This sequence follows the rule: "Never pass on a defect and never accept a defect." Such closeness between producer and internal customer has enabled Toyota plants to come nearer to producing defect-free vehicles than any other factories in the world.

- The new customer-driven approach to manufacturing has led to a customer-driven approach to machinery. Traditionally, manufacturers designed machines for speed. But the customer-driven approach has caused Toyota and other manufacturers to realize that higher priorities were flexibility and ability to prevent defects. Ohno calls it "automation with a human touch." Toyota created devices that would stop a machine if a part wasn't right. With years of experience, Toyota can now make these machines run as fast as or even faster than conventional machines, and the customer is served better.

- The system also eliminates deterioration of parts in storage, saves space, and saves the interest companies would have to pay to finance larger inventories.

Breaking down the barriers that force employees to work in complicated batch-oriented systems can vastly improve service companies as well. How did MBNA America, the large credit card company discussed in the last chapter, reduce the time required to approve a credit-line increase from the industry-standard eight days to just *one hour*? By eliminating all the batch-oriented procedures that other banks use. Today's computer technology makes it possible for companies to eliminate many complex batch processes (by using on-line credit reports instead of telephone calls, for instance). On the other hand, it can also cause new setup-time barriers, as when employees must load special software to do a task.

■ Smashing Barriers to Innovation

Customer needs must also drive the development of technology. Many companies today create technical and design innovations far faster and with greater attention to customer needs than ever before.

Xerox has found that by analyzing the *process* of innovation and breaking down barriers, it can develop products two-thirds faster and cheaper than a decade ago.[9]

Xerox's analysis of the old process of innovation showed that:

- No one had been clearly in charge of development projects from start to finish. The solution in the new system for innovation: designate a chief engineer in charge of the project from concept to start of manufacturing.
- Engineering managers too readily approved the basic design of machines without feedback from the manufacturing and marketing

departments or from the specialists who would design components such as the paper-feed trays. As a result, designs often had to be radically reworked at later stages. The solution in the new process of innovation: spend longer on the initial "conceptual design." Before a lot of money is spent on designing individual components, get suggestions from dozens of groups who will produce and market the product. Do this before the design gets its first approvals.

- Barriers among engineering, marketing, and production departments were slowing development. When a project reached a new stage that required participation by marketing and production people, its progress would halt while the new people were taught details the engineers had long ago mastered. The solution in the new procedure for innovation: Involve marketing and design people at earlier stages in the design process.

The result is a far more customer-focused project that ultimately takes much less time because work doesn't have to be redone.

Moreover, companies like Xerox also rigorously analyze each process that contributes to the innovation effort, seeking better, more customer-driven ways of working. There's a careful eye for improvement in each of these processes:

- learning customers' needs,
- keeping up with technology outside the firm,
- maintaining basic technical strengths,
- discovering new technology that will be useful in products,
- neutralizing competitors' basic technical strengths, and
- keeping the organization as a whole informed about discoveries.

Once barriers are dismantled, innovation often makes a startling forward leap. United Technologies Corporation's Automotive Division, which makes various auto parts, cut the average time needed to respond to a new request from an automaker from six months to less than six weeks.

■ Create Customer-Friendly Policies

Some organizations have a special knack for infuriating their customers. These companies turn them down and off with the maddeningly dead-ended statement: "It's our policy."

Reaction? Most customers feel just like you did when you were nine years old and your mother refused you something, saying: "Because I say so."

To be sure, people who work with customers have to follow some standard policies. But those policies don't have to become barriers to customer satisfaction. Unfriendly and often useless rules and regulations promulgated by bureaucrats are among the great impediments to creating happy customers.

This chart shows four kinds of policies.

The Policy Matrix

	Customer friendly	Not customer friendly
Necessary	We accept credit cards	Insurance regulations prohibit customers in parts of the shop area
Unnecessary	We'll take back merchandise, no questions asked	We do not accept returned merchandise

The principles for handling these policies are summarized in this chart.

	Customer friendly	Not customer friendly
Necessary	KEEP THESE— CUSTOMERS EXPECT THEM	1. **Challenge** to see if really necessary, eliminate if not. If you can't, then 2. **Change** to make friendlier and **move** to the customer-friendly side. If you can't, then 3. **Train** employees to give customers a proper rationale, something more helpful than, "It's our policy."
Unnecessary	CREATE MORE— THEY CREATE LOYALTY	GET RID OF THESE—THEY'RE COSTING YOU CUSTOMERS

Review the policies of your company, department, or business unit. The first goal is to eliminate all policies that are "unnecessary" and "not customer friendly." Often old, out-moded, customer-irritating policies remain in force for no other reason than "It's the way we've always done it." Get rid of them.

Second, examine the "unfriendly" but "necessary" policies. Challenge them. Are they really necessary? Ask yourself: What would be the consequences of eliminating this policy? Would the resulting problems outweigh the aggravation we're causing our customers by retaining the policies?

If an unfriendly policy must stay, can you change it to make it friendlier? For example:

Unfriendly	Friendly
We do not credit a check to an account until five days after it is deposited *becomes*	We can credit checks immediately on request for depositors in good standing for more than one year.
All loans will require four approvals and department-head signatures *becomes*	Loans for less than $5,000 will be approved by the loan officer handling the transaction.

Ultimately some "customer unfriendly" policies will remain for legal, insurance, safety, and economic reasons. Carefully review these to analyze whether the same goals can be accomplished in a more customer-friendly way. For example, if fears of liability keep a customer from being in the shop where his or her car is being fixed, can the customer wait and meet with the mechanic who worked on the car somewhere where they can at least see the shop area? If you can't achieve this get-together, train your people to explain the policies in a friendly, understanding way.

The customer-friendly side is easy. Keep the ones that are necessary and create more "unnecessary" ones. The "unnecessary" policies are the ones that delight your customers because they don't expect them. These are gold, the stuff of which competitive advantage is made.

■ BREAKING DOWN BARRIERS
TO SERVICE

In general, service businesses haven't collapsed barriers with the zeal and acumen that a handful of great manufacturers have displayed. Manufacturers' service departments have often been laggards, too. There is, of course, no reason why such customer-driven efficiency can't be achieved, and today some service organizations are learning to excel at just that.

L. L. Bean carefully manages every aspect of the customer's experience. It measures and tracks its performance in widely varied "key result areas" (KRAs), including seven in customer service and others dealing with Bean products, personnel, and so on. The seven KRAs for customer service are:

- convenience,
- product guarantee,
- in-stock availability,
- fulfillment time,
- innovation,
- image, and
- retail service (for the Bean retail store in Freeport, Maine).

Under the heading of convenience, for example, Bean tracks how long it takes telephone customers to reach the company (85 to 90 percent should reach an operator or a recorded message within three rings), how many calls are abandoned by the caller (it should be fewer than 2 percent), and so on. Bean also benchmarks its performance against about eight competing mail-order houses. Bean people place orders elsewhere and compare performance to that which the KRA system has shown to be typical at Bean.[10]

The barriers to excellence in service, of course, often differ from those in manufacturing. In factories, the biggest barrier that must be smashed is that manufacturing systems cut people off from the customer. They frequently limit people's interaction with the internal customer who will use their work, and they completely cut off interaction with the ultimate user.

A few service organizations—a water company, for instance, or the check-processing group of a bank—may suffer from exactly the same kinds of barriers to serving the customer as a manufacturer.

But in general, the barriers to quality of service are subtly different. Service people must often "produce" and deliver their product in widely scattered locations, so that it's easy for new barriers to serving the customer to grow into their operations. Moreover, because so many service employees deal directly with the customer, excellence in service depends even more than excellence in manufacturing on helping employees to care for the customer deeply throughout what can be an emotionally exhausting day. And the very closeness of customers, who seem able to create an infinite number of unpredictable situations, is a barrier to setting the high standards necessary to excel. Managers can easily blame the customer or an individual employee for the organization's failures. It's even easier for service organizations to find excuses for failure than for manufacturers.

But when companies establish measures like Bean's KRA standards, and when they carefully examine why performance isn't what they'd like, they soon realize that the differences between manufacturing and service are much greater in perception than in reality. Customers have unhappy experiences because:

- employees don't know they're supposed to create happy experiences,
- employees don't get the resources they need to accomplish this service,
- employees lack the training to use the resources they have,
- employees have incentives that don't encourage them to work for customers.

■ Constantly Look for More Barriers to Break

The best of the practice that many companies call Total Quality Control is eliminating barriers to serving the customer. But if you focus on breaking down the barriers to serving the customer, rather than the vaguely defined word "quality," you'll solve problems you'd otherwise miss:

- **Barriers to serving the customer in marketing.** *Sales and Marketing Magazine* recently said, "The best way to sell is to make it easy to buy." Perhaps the most destructive barriers to customer satisfaction are the ones that keep customers who need your product from buying it. Yet sometimes we're so busy removing barriers to high quality in our factories and in our back offices that we don't think about the barriers we've set up that make it hard to buy. Are your offices hard to find? Do credit approvals take inordinate time? Do you require unnecessarily large minimum orders? Are your salespeople invisible when a customer wants to buy?

- **Billing problems.** It's amazing how many organizations can't send an invoice out properly. Sometimes these bills are simply inaccurate. Sometimes they're merely incomprehensible. Either way, companies have a big opportunity to create a billing procedure that doesn't raise barriers.

- **Empathy that entraps.** When a customer calls in because the copier is broken and your staff can't get there till 2:30, the customer is mad. The cry is: "Can't you get here earlier?" And your responsive service rep is likely to say, "Okay, I'll try. Maybe we can be there before 2:00." Naturally your customer will be highly dissatisfied if the serviceperson walks in at 2:30. Some companies are attacking this pitfall. When you call Xerox for service, the staff is asked to project a time 20 percent later than the time it actually believes a serviceperson can be there. If you're a customer and you're told a serviceperson will be there at 3:00, how do you think you'll feel when he or she walks in at 2:30? Less angry? Sure; maybe even pleased. The rule here is: *underpromise and overdeliver.*

Even L. L. Bean, a very good company, hasn't yet solved all its problems. In fact, Bean applied for the Baldrige National Quality Award and, like other service-company applicants, didn't get it. The company was concerned to discover that judges from leading manufacturing companies felt L. L. Bean too often has to fix customers' problems rather than doing things right the first time. As a result of the ensuing soul-searching, L. L. Bean is deliberately cutting back its growth while it improves its systems still further.

Imperfect or not, L. L. Bean has already shown that every aspect of a service business can be managed zealously. It has significantly lower return rates than competitors, and until it cut back growth, sales were expanding about 50 percent faster than those of the mail-order industry as a whole.

Other service companies have similar systems for managing cleanliness of hotel rooms, how long bank patrons must stand in line, or how speedily power is restored after being knocked out by lightning. Most haven't quite equaled the best manufacturers, but they've shown that any procedure can be managed to serve the customer dependably.

■ TOP MANAGEMENT AND THE BARRIER-FREE COMPANY

Most barriers develop accidentally. They're often by-products of some-one's honest effort to improve the company. And so leadership in barrier-smashing is difficult. Top managers can easily set out to eliminate barriers and merely wind up creating new bureaucracies. When the U.S. Congress passed the so-called Paperwork Reduction Act, one of the major results was the creation of regulations requiring "Paperwork Reduction Act Notices." You guessed it—more paperwork.

But it doesn't have to be that way. Organizations can take big steps toward eliminating barriers by changing their way of managing them-selves. Four management strategies can make the difference:

1. **Deriving the organization's official policies from its vision** and its day-to-day management practices from those policies.

2. **Creating cross-functional management structures** to mobilize people from all parts of the company in solving key problems.

3. **Adopting "visible management"** to help managers and other employees recognize customers' needs and monitor progress toward meeting those needs.

4. **Carrying out regular presidential reviews of the processes and control points** used in managing each part of the business.

Let's look at each of these barrier-bashing top-management policies in turn.

■ Aligning to a Vision Leads to a Barrier-Free Business

The most fundamental element of an organization is its vision statement because it represents what that organization strives to be. In essence, a vision statement represents a group's or individual's reason for being in the organization.

Insist that policies be aligned with the vision and you take a fundamental step toward aligning yourself with the customer. You take the most important step toward creating organizations that instead of creating barriers to customer satisfaction break those barriers down.

Case in point: the way in which Ray Kroc, founder of McDonald's, ran his company. Kroc had a vision for McDonald's and it was clearly stated: "Quality, Service, Cleanliness, Value." He drove everything his company did according to policies based on that vision. Whether it was store design or toasting hamburger buns, Kroc had a policy and the policy was aimed at giving the customer Quality, Service, Cleanliness, and Value. "When Ray talked about McDonald's, he could touch people, move them," recalled Fred Turner, a former counterman in the Oak Bluffs, Illinois, McDonald's and now McDonald's chairman. "When he talked about the bun and how you toast it, you could see the bun. By the time he got through talking about buns, you were hungry."[11]

It's also a fact that success—or perhaps more accurately, the complacency that comes with success—can be a barrier, too.

When TGI Friday's opened its first seven restaurants in 1975, business was so good that customers were lined up outside the doors. But not long after that, whoosh: sales dropped 50 percent.

Daniel R. Scoggin, who was chief of TGI Friday's, recalls that management was baffled for a while, and then finally took a hard look at the restaurants themselves. Customers were being turned off; their expectations were not being realized. Management found thousands of tiny problems: signs hanging crookedly, untidy restrooms, waiters who weren't as friendly as they could have been. Of course, TGI Friday's wanted to be casual, but it had gotten sloppy.

The company established procedures to make sure those tiny terrors would be taken care of, and sales quickly recovered. Friday's established a policy that every manager of every restaurant had to sit in every seat in that restaurant at least once a month. In effect, he or she would walk in a customer's shoes (or sit in the customer's seat, at least) often—very often.

"Whenever a manager told us sales were down because of competition or the local economy or other outside factors," Scoggins says, "we took a hard look inside. We drew up a checklist of problems that needed correcting inside the restaurant—everything from a waiter's bad attitude to a burned-out lightbulb. By the time the manager got to the end of the list, sales had rebounded."[12]

When a company is aligned with its vision, it has a management structure that each year requires people in all parts of the organization to identify the most harmful barriers to giving the customer what the customer wants. And it insists that each year every manager remove the most significant barriers in his or her part of the company.

■ Unite People Across Functions

Can companies ever be united organisms? Many consider corporate infighting an inevitable consequence of bigness. The finance department screams about expense accounts, marketing complains about production failures, the factory denounces the design engineers, and the folks in research and development (R&D) look down on everyone. At one major United States bank, a marketing group trying to improve relations with customers had to fight a guerrilla war just to get the customers' names from bank lending officers. The barriers between people carrying out different functions prevent them from doing all they could to serve customers.

Today, companies that fail to stamp out such internecine warfare—hostility complicated by myopia—will not succeed; that's understood. To compete, organizations have to function as united teams that break down, not create, barriers.

A few companies have developed systems of **cross-functional management** that show how we can tap all our people's abilities to attain our goals. Though department heads still have charge of their departments, marketing is no longer the sole province of marketing people or production the private territory of production people.

Instead, a committee of senior managers provides guidance in the areas crucial to the firm's long-term growth. The committee includes both top managers who work in these areas full time *and* other equally qualified personnel whose principal jobs involve different areas. The chairman of the quality committee, for instance, is an executive whose principal duties are in achieving top quality. But the other members may include the head of a factory and a marketing executive.

One manufacturing company has such cross-functional committees in charge of:

- quality,
- costs,
- technology,

- production,
- marketing, and
- personnel.[13]

Such top-management committees, because they *are* in top management, can achieve more cross-department influence than ordinary staff sections ever could by lobbying and cajoling. They break down barriers between sections. A top-ranking marketing manager, say, can work with the quality and production executives on the quality committee to make sure that marketing people provide the information production and design people need. What's more, when they are working well they create an excellent example of cross-functional collaboration for all at lower levels in the organization to emulate.

The Japanese have a useful saying:

To move information, move people.

The cross-functional top-management committees can ensure that enough people are reassigned between research and development and the production departments, or between the production department and the sales department, the each group has a reasonably clear understanding of the other.

Cross-functional management also achieves dramatic improvements at lower levels. Until recently Honeywell took four years to design new thermostats. But when a customer threatened to take business elsewhere, Honeywell set up a "tiger team" of marketing, design, and engineering employees and gave it total freedom. The team broke all Honeywell's traditional rules, and, probably because of that, the job got done in twelve months.[14]

■ Making the Company's Struggle Visible

To keep goals and problems constantly before them, well-managed firms use an array of techniques designed to make everything going on in the company *visible*.

In a Komatsu factory the company vision is posted at the entrance to the plant. Work standards for each job are posted at the job site so that anyone can see whether the worker is doing it right or not. Lines are painted on the floor exactly where inventory boxes and trash cans should be placed. Charts show the number of suggestions made by each group and often each individual. Charts track downtime, process variations, measures of customer satisfaction, and safety.

Assembly operations use a "kit control system"—parts arrive in a bucket with the exact number of screws and other parts needed for assembly. If anything is left over, the worker will see that something has been forgotten.

Ordinary employees make the charts themselves; typically, the person whose performance is being measured draws the graph. The workplace may look "unprofessional" by Western standards—more like a school before a pep rally than like a high-technology factory. But by involving everyone in visual management, Komatsu builds everyone's commitment to the company's objectives and each worker's part in achieving them.

In most companies, especially large ones, people lack knowledge about what's going on a few feet away. That becomes one of the biggest barriers to a unified team effort. Make what's happening *visible*, and you'll ensure it's on people's minds.

■ Controlling Barrier-Breaking Through a Presidential Review

A chief executive can do more than anyone else to ensure barrier-breaking by reviewing the operation of his or her company's business units—not so much to study their achievement of objectives, as to ensure that they're managing procedures well. Annually in each part of the company the leader can ask:

- Has management adopted policies that will achieve the organization's vision?
- Does management have a clear plan for improving procedures to eliminate barriers? Is this year's plan for removing barriers aggressive enough, and is it being accomplished?
- Is management measuring and controlling the right things?
- Is management putting its data into visual form?

In this way, the leader can learn the underlying realities in the workplace. And the people in the company will know that the whole organization is serious about its goals.

■ The Barrier-Free Company

Tony Clarry, senior general manager of customer service at British Airways, told me:

> We tell our front-line people: You're responsible for creating a happy customer. If you have a problem, solve it. If you can't solve it, give it to your supervisor. If your supervisor can't solve it, he or she passes it right up the line.

That's a good way of looking at our jobs. Sir Colin Marshall, chief executive at British Airways, had on his desk the problems nobody could solve that were getting in the way of serving customers. The problems went right up the line, and he solved the few that didn't get handled along the way.

It really can work. When you break down the barriers to serving the customer, you're creating an organization that will truly focus on the customer—and win.

ACTION POINTS

- Don't wait. Begin programs to solicit ideas from employees, suppliers, and distributors on breaking down the barriers that prevent you from doing right by the customer.
- Starting with managers, teach everyone in your organization basic techniques for prioritizing problems and solving them to serve the customer.
- Develop communications channels so that everyone in your organization will know about the barriers that your customer research and benchmarking activities reveal.
- Does your business involve complex processes like manufacturing a highly technical product or processing complicated paperwork? If so, review whether long setup times for the parts of the process are creating a batch-oriented system that can't be driven by customers' needs. If the answer is yes, drive down setup times and reorganize the process so that it serves the customer.
- Analyze your innovation process. Does it smoothly serve the customer? If not, reorganize it so that it does.
- Create top-management structures that will break down turf barriers. They prevent you from fulfilling customers' wants.
- Adopt "visible-management" systems that enable everyone to see your goals and your progress. Such systems also make visible the barriers that are blocking you from your goals.

■ RESOURCES

H. James Harrington, *The Improvement Process* (Milwaukee, Wis.: American Society for Quality Control/Quality Press, 1987).

Masaaki Imai, *Kaizen* (New York: Random House Business Division, 1986).

Kaoru Ishikawa, *What Is Total Quality Control? The Japanese Way* (Englewood Cliffs, N.J.: Prentice-Hall, 1985).

Taiichi Ohno, *Toyota Production System: Beyond Large-Scale Production* (Cambridge, Mass.: Productivity Press, 1988) [Japanese edition 1978].

J. M. Juran, *Juran on Leadership for Quality* (New York: Free Press, 1989).

■ NOTES

1. *Japanese Manufacturing Techniques: Nine Hidden Lessons in Simplicity* (New York: Free Press, 1982), p. 103.

2. Sidney (Shuichi) Yoshida, "Quality Improvement and TQC Management at Calsonic in Japan and Overseas," paper prepared for the Second International Quality Symposium in Mexico, November 1989.

3. Calculated from *Fortune* magazine Industrial 500 and International 500 directories, published April 24, 1989 and July 31, 1989. The comparison is inexact because the two companies have different methods for consolidating results of subsidiaries, and so on, but it is still a significant indicator of relative productivity. Toyota achieves nearly four times as many sales per employee as General Motors, but that comparison is misleading because General Motors makes a larger share of its own parts than either Toyota or Chrysler.

4. Alton E. Doody and Ron Bingamin, *Reinventing the Wheels* (Cambridge, Mass.: Ballinger, 1988), pp. 77–82.

5. Doody and Bingamin, p. 97.

6. *Tom Peters' On Achieving Excellence,* 3:3 (March 1988).

7. Taiichi Ohno, *Toyota Production System* (Cambridge, Mass.: Productivity Press, 1988).

8. Maryann Keller, *Rude Awakening* (New York: Morrow, 1989), pp. 211–213.

9. Robert Chapman Wood, "Quality by Design," *Quality Review* (Spring 1988), pp. 23–27.

10. Bernard J. LaLonde et al., *Customer Service: A Management Perspective* (Oak Brook, Ill.: Council of Logistics Management, 1988), pp. 117/ff.

11. John F. Love, *McDonald's: Behind the Arches* (New York: Bantam, 1986), provides an outstanding, well-documented account of the McDonald's Corporation's commitment to the customer.

12. Daniel R. Scoggin, "Customers Go Out the Door When Success Goes to Your Head," *Wall Street Journal,* August 11, 1986.

13. Kaoru Ishikawa, *What Is Total Quality Control?* (Englewood Cliffs, N.J.: Prentice-Hall, 1985), p. 113/ff. Ishikawa invented much of the cross-function management systems now used in Japan, and the entire discussion in that book (pp. 113–118) is extremely useful.

14. John Bussey and Douglas R. Sease, "Speeding Up: Manufacturers Strive to Slice Time Needed to Develop Products," *Wall Street Journal,* February 23, 1988.

The *Quality measure is customer opinion. In the
past we spent a lot of time telling customers they
were wrong. We can't do that any more.*
— Joel Ramich, Associate
Director of Quality,
Corning Works

6

Measure, Measure, Measure

Water reeking of sulphur was the bane of Joban Kosan, a Japanese coal-mining company facing extinction in the 1960s. The water bubbled under much of the land the company owned, filling mine shafts and creating turgid yellow pools. Coal itself was no longer a profitable mineral, and Joban, on the brink of going out of business, wanted to use its huge acreage for other purposes. But what about the sulphur-tainted water? Would the smell of sulphur drive other users away from the land as well? And what about the coal mine's employees?

Then someone had a stroke of genius. Don't get rid of the water, use it. Today Joban is renowned, not for its coal mines (long abandoned) but for a pleasant resort featuring, yes, hot sulphur springs.

Joban Spa Resort Hawaiian, a few hours from Tokyo, is a very unusual playground. It is the biggest moneymaker for the now-diversified Joban

Kosan, and it practices a very unusual brand of "fact-based management." It profoundly values and constantly uses information obtained from a thorough program of detailed measurement, that, as far as I know, is unique in the service industry. The spa's general manager is a former coal miner; the person responsible for quality came from an affiliated foundry. What could they know about running a resort? Plenty, I found out.

I visited Joban Spa Resort Hawaiian in the mountains of Honshu in 1990. And I soon realized that Joban knows (1) how to learn what guests want, and (2) how to give it to them. In fact, in 1989 the resort, which caters to middle-class Japanese, became the first leisure-industry company to win the prestigious Deming Prize for Quality Control.

Susumu Yamada is head of the Spa's Quality Promotion office and a former quality manager of a Joban subsidiary that makes metal castings. He believes the procedure for learning what customers want and then giving it to them at a resort hotel is almost exactly like learning what customers want and giving it to them at the metal foundry. "Basically, there's no difference," he says.

Joban quality experts distinguish between "physical" quality and "mental" quality. "Physical" quality is essentially what I have called "product quality"—the "What you get." "Mental" quality is essentially what I have called "service quality"—the "How you get it." Yamada says service industries have to put "more emphasis on mental quality." But otherwise the procedure for doing right by customers is amazingly similar in a manufacturing company and in a service company.

Joban measures everything:

- It has videotaped the faces of people standing in check-in lines to determine how long they'll wait before becoming annoyed. Then it has established standards and procedures that prevent them from ever having to wait that long.
- It has thoroughly analyzed the level and type of background music that customers enjoy, and it carefully provides exactly that.

- It has created Quality Circles—small problem-solving groups— among everyone from front-desk clerks to dancers in the nightclub. It has taught them all to measure every action to ensure that they give customers what they want. And they do so.

Joban starts its measurement system with the customer's needs and works back. It simply figures out, from the customer's desires, exactly what should be measured at each step in the method of doing business. Admittedly, not every customer's desire can be quantified. Even customers themselves don't know everything they want. A dance troupe's Quality Circle can never choreograph a great work of art just by measuring customers' opinions. But by making customers' needs drive its measurement program, Joban captures a clear understanding of what matters to customers, and captures such magnificent benefits that the former coal-mining company now offers lessons to the rest of the world.

In the 1990's, organizations can neither build nor keep market share unless they learn to measure their successes and failures from the customer's point of view. Companies like Joban Kosan—and also much-better-known firms such as Minnesota Mining and Manufacturing (3M) and Apple Computer—have prospered by *both* rigorously quantifying customers' desires that can be quantified *and* helping their people innovate to address the desires that can't.

In this chapter we look closely at measurement—something that most companies do, but few do well. We're going to show that you *can* measure well, using measurement to deliver more of what the customer wants, if you follow five principles:

1. **Know why** you're measuring.
2. **Let customers** tell you which end results to measure.
3. **Ask continuously** how well you—and your competitors—are doing.
4. **Track the internal procedures** that are supposed to produce the results the customers tell you they want—as well as the end results.
5. **Tell your people** everything you learn.

■ KNOW WHY YOU'RE MEASURING

All measurement is an opportunity to focus on and meet your customers' needs. But most companies don't use it in that way.

If I tell a group of businesspeople that they should constantly measure how their customers see them, nine times out of ten they'll say they're already doing it. Then I ask them: "How well are you doing with your customers right now? And how do you know?"

A few give me good answers. If they can tell me. . .

> We survey our customers every month, and we've improved on twelve of our fourteen measures of customer satisfaction this year. And I have a lot of confidence in those measures because we did a series of focus groups last year to make sure we were measuring the right things. . . .

then I believe they have a customer-focused measuring program.

More often, businesspeople respond by talking about extremely indirect, unreliable measurements. They say things like:

> "Our market share is increasing."
> "Orders are ahead of last year."
> "We aren't getting any complaints."
> "We have a good market-research group and they say we're in tune with our customers."

When I hear these comments, I have a strong sense that the people making them don't really know *what* the customer thinks. Market share and the volume of new orders are good indicators of customers' opinions, but they are *lagging* indicators. By the time you learn of a problem from market-share statistics, your competitor may have seized an advantage over you that's so big you'll never recover. Moreover, many companies get "no complaints" simply because their customers think the companies are not interested in listening. In a survey of 2,000 American consumers for the *Wall Street Journal*, only 5 percent reported that they thought American business was listening to them and striving to do its best. And market researchers too frequently measure only a small part of the customer opinion you need to understand.

Measurement programs should establish clear, quantitative channels that will communicate everything that is important to the customer to the whole organization. They should be structured so that they constantly help *every* part of the organization to serve the customer better.

Well-designed measurement programs have a great many benefits:

1. Measuring what the customer wants and whether you're providing it forces a discipline that will benefit your organization even before you learn the results. It forces you to think about critical success factors and main elements in achieving these factors. It demands that people in the organization discuss what is important and deal with disagreements.

2. Good measurement systems provide reliable information on what's right and wrong with the organization. Progress (or lack or progress) against measures tells people how they're doing and helps them improve.

3. A good measurement system provides an agreed-upon framework in which people can discuss the organization's procedures and problems.

4. Measurement systems accurately tell you what you do well, so that you can promote yourself honestly in the marketplace.

5. People derive justifiable pride by achieving and exceeding their goals on the measures. The measurement system serves as a basis for recognizing outstanding performers—and provides information you need to help sub-par performers improve their results.

Good measures can also help you create well-designed incentive compensation. But be careful: Whenever someone's compensation is based on specific numbers, the driving incentive that is created may be to manipulate those numbers rather than to improve service to customers.

■ How Measurement Systems Go Wrong

Most businesses never achieve good measurement. Some business-measurement systems are designed not so much to help companies serve customers as for public relations, legal, regulatory, and advertising purposes. The people carrying out the measurement know what managers want to hear and are influenced by that knowledge. The managers may have no way to judge whether the data they're getting accurately reflect customers' real feelings and needs.

Other companies do market-research measurement for purposes that are good, but not good enough. Some companies study customers' favorite colors. (Will people buy more Acme detergent if we dye it lime green?) They study demographics. (If children of the baby-boom years are growing into middle age, should we introduce a new wrinkle cream?) They interview customers to ask what new products those customers would be willing to buy.

These studies often *do* result in better, more useful products. But they don't give today's companies the comprehensive understanding of their relationships with customers that they need.

Most market researchers ask about only a few aspects of the customer and the product—aspects that, if altered, might produce a quick boost in sales. Worse, companies usually communicate the information only to a handful of product-development and marketing professionals. Such half measures don't link the organization as a whole to its customers.

True customer-driven research, on the other hand, is designed from the beginning to engage the entire organization. It has inherent power to change the way in which the organization thinks, and then improve the organization again and again in a refinement driven by customers' needs. Customer-focused measurement inevitably sticks its nose into every part of the company—a highly desirable nosiness.

If this is your purpose, don't limit your focus to questions about whether customers would buy a purple box more readily than a blue one. Rather, start with the truly vital questions about the relationship between your company and its customers. Prepare for painful answers, but also invaluable insights. Customer-keeping research helps you to know:

How do you like doing business with us? How can we change for the better?

This research leads to a system that will monitor all the events that contribute to customers' satisfaction. It demands communication of the data you collect from the customer to every person in the organization.

When British Airways created a video complaint booth in Heathrow Airport, it initially showed the tapes of angry customers only to top executives. Happily, though, British Airways was in the midst of a massive program to improve its London operations, and managers were spending plenty of time with the front-line people. They soon heard a clamor. The people on the front line wanted to see the tapes. That's when the program began to have dramatic effects on the organization.

Most people in every organization want to create happy customers. And most people want accurate information about how well they're doing. But people can't satisfy customers consistently unless their company measures its performance thoroughly. And people won't know how to improve unless their organization gives them information.

■ LET YOUR CUSTOMERS TELL YOU WHAT TO MEASURE

The chart on page 156 summarizes the specific information you need to learn from your customers and how you can obtain it. For this summary we assume that you've already taken the basic steps we discussed in Chapter 2:

- Defining who your customers are;
- Learning to listen to their voices.

A customer-focused measurement program starts with open-minded listening. In this part of the chapter, we'll discuss the first three questions in the chart:

Type of information	Methods for obtaining
Which product and service characteristics may be important to your customers?	Focus groups, other methods for asking "open" questions
What is the relative importance of each characteristic?	Customer surveys, other methods
What level of performance on each characteristic is satisfactory to customers?	Customer surveys, other methods
How well are you actually providing each product and service characteristic?	Customer surveys, "mystery shoppers"
Are your internal procedures under control, and are they performing in a way that will create the product and service characteristics most important to customers consistently and with constant improvement?	Internal process measures (defect rates, statistical measures of variation, "diagnosis" of the functioning of each unit within the organization, and so on)

- Which product and service **characteristics** are important to your customers?
- What is the **relative** importance of each of their wants?
- What **level** of performance on each product and service characteristic will meet the customers' expectations?

The answers to these questions will tell you, in quantitative form, what **matters** to customers. Once you know that, you'll be able to create a system to track how well you're satisfying them day by day and month by month. And you'll be able to create internal measures that consistently inform your people how successfully each of them is contributing to the product and service characteristics that customers want.

■ Which Product and Service Characteristics May Be Important?

Measurement is worthless if the achievements being measured fail to benefit the customer, and too many of the measures in today's corporations fail in exactly that way. Often the designers of the measurement program didn't even consider some of the product and service characteristics that were actually most important to the customers.

When one auto importer launched its cars in the United States, its service organization had what it thought was an extremely advanced computerized system for measuring performance. But this measurement initially hurt the customer at least as much as it helped.

The company's system, like many others, was focused on a few performance indicators. The service organization focused on the share of dealers' requests for parts it succeeded in filling overnight. Achieving a fulfillment rate of 90 percent seemed a good performance.

In other words, when a dealer ordered a part, it got the part it wanted the next day from a regional warehouse nine times out of ten. But when the company asked customers themselves what *they* thought about availability of replacement parts, the majority said they had experienced significant delays.

The company's measurements weren't inaccurate, but the company wasn't asking about the service characteristics most important to the customer. Sure, nine out of ten times when a dealer ordered a part, it would arrive the next day. The problem was: What happened that tenth time? The company didn't track whether the part ordered was needed to repair a car sitting inoperable in the dealer's lot or was merely destined for the dealer's inventory. Unless a dealer specified overnight express shipment, days or weeks might pass before a part that couldn't be delivered overnight reached the dealer. And dealers avoided specifying overnight express shipment except in extreme cases because they disliked paying the cost. *Sixty-seven percent of customers who had needed parts had experienced delays of two weeks or more in fulfillment of a parts request.*

This firm's measurement system was focused on the overnight order-fulfillment *percentage.* But the 10 percent failure rate meant that most customers experienced at least some delay in getting parts. And the system obscured the long waits suffered on relatively few orders, waits that ultimately affected 67 percent of parts customers.

But the company was also developing an advanced system for measuring customers' satisfaction. That system warned the company about dissatisfaction with delivery of parts. After carefully reviewing the service characteristics that really affected customers, the firm overhauled its service organization's measurement system. Still tracking the "fulfillment rate," it also began tracking the number of dealers' complaints about availability of parts. And it began managing the business to satisfy customers. It made overnight express charges for back-ordered parts an allowable expense on warranty work, and created an improved computer tracking system so that central offices would know which dealers had which parts in stock. Dealers created a system of "negotiated splits," allowing a dealer in one town who needed a part to obtain the part from a dealer in the next town in return for a portion of the profit from the repair. The result: increased customer satisfaction—all because the company turned to its customers to learn what it should be measuring.

■ How to Listen to Customers at the Start of a Measurement Program

To avoid problems like this one, take time at the start of a measurement program to develop a comprehensive list of the characteristics and problems that may be significant to your customers. You can't meaningfully measure (or even guess) what's most important until you have created a comprehensive list of product and service characteristics that *may* be important.

Developing that list requires open communication—living close to your customers so that you know their needs. Joban Spa Resort Hawaiian is expert at this kind of listening: It not only interviews people to learn what they want, it also discreetly follows randomly selected guests around the resort to record what they seem to appreciate and what they avoid.

Start by asking yourself, "What characteristics are my customers *likely* to be seeking in my product and service?" Make a list, then add to it by using techniques discussed in Chapter 2, such as:

- **"Focus Groups."** Gather a small group of customers and ask about their problems and expectations.
- **Investing in complaints.** Do your best to encourage people to complain, then list the issues they complain about.
- **Open-ended questions on customer comment cards** and in customer interview surveys.
- **Visits to customers.** If you make a consumer product, visit your customers in their homes. If you make a product for business, visit them at their company offices.
- **Surveys about customers' reactions to competitors' offerings.**

In developing lists of characteristics for services, consider the RATER checklist (Reliability, Assurance, Tangibles, Empathy, Responsiveness) introduced in Chapter 2 (page 47). One United States publishing company carried out an analysis like this in its order-fulfillment and shipping departments. Under the heading Reliability, the report listed these quality characteristics:

Accuracy of description
- Items match their descriptions and pictures in catalogue
- Catalogue advertising is truthful.

Performance of product
- Product does what it's supposed to.

Record keeping
- Accurate records are kept
- Records are updated promptly.
- Records are used to prevent errors in billing and other functions.

Accuracy of Shipment
- The proper merchandise is shipped.
- Nothing is missing from shipments and all parts are enclosed.
- Products are not substituted without permission.

- Back orders are avoided.
- Reliable means of delivery are used.

Warranty-Guarantee Performance

- Warranties and guarantees are honored.

The company made a similar list of quality characteristics for each of the other aspects of service performance: assurance, tangibles, empathy, and responsiveness, and also value. A larger list of qualities such as that on page 253 can be useful in preparing lists that will cover *both* manufacturing *and* service.

■ Quantitative Questions: What Level and How Important?

Once you have a list of product and service characteristics, seek answers to the quantitative questions crucial to customer-focused measurement:

- What is the relative importance of each characteristic to the customer?
- What level of performance on each characteristic is satisfactory?

You'll need to find a unit of measure for each desire (see the accompanying table).

Customer expectation	Unit of measure
No unreasonable wait to check in.	Minutes on line
The delivery is on time.	Minutes, hours, days
The computer system stays on-line.	Percentage of "up time"
No interruption in service.	Percentage of "down time"
A durable product.	Hours of use before breakdown
Beauty of finish.	Number of blemishes

When you can't capture the customer's desire in a numerical unit, you'll have to ask for subjective impressions on a scale such as this:

Poor   Satisfactory   Excellent

1. . . . . 2. . . . . 3. . . . . 4. . . . . 5. . . . . 6. 7

Customer expectation	Measurement
Service people show empathy toward customers	Response of customers to a question such as: "On a scale of 1 to 7, please rate how well the people of XYZ company provided caring individual attention to you as a customer."
Friendly telephone operators	Response of customers to a question such as: "On a scale of 1 to 7, please rate our telephone operators for friendliness."

And some customer expectations may call for measurement in several units. "Beauty of finish" may be measured not only by number of blemishes, but also by amount of light reflected or purity of color. Use enough measures so that you're sure you'll capture what the customer really wants.

Once you have a thorough list of characteristics and appropriate units of measure for each, the easy way to answer the quantitative questions is simply to ask your customers the questions, in interviews and written questionnaires. If you run a small company or a small workgroup within a larger organization, that's usually all you have to do.

Answers obtained in this way, however, may not be accurate. Suppose Joban Spa Resort Hawaiian had asked customers how long a wait to check in was reasonable. Many of them might answer: "No time." In truth, they probably wouldn't know. What's reasonable, anyway?

Joban's decision to videotape customers in line therefore made good sense. It gave Joban real knowledge of how much waiting causes annoyance, and today when lines near that length, additional Joban staff members are brought to the check-in desk to reduce them.

There's one consistently reliable method for learning what *really* satisfies people and the *real* relative importance of each customer desire. You can carry out statistical analyses that correlate customers' actual experiences with the same customers' reactions to those experiences. Interview customers just after they've completed an experience with your product—an airline flight, say, or the first two years of owning a manufactured good. For each customer, learn how your organization performed on each of the product or service characteristics on your list. And ask each to tell you how he or she feels about the experience. Statistical analysis of these data can then tell you the levels of performance on each of the characteristics that correlate with high satisfaction and the relative strength of the correlations for each of the characteristics. That will give you the clearest possible indication of the relative importance of each characteristic and the level of performance on each that's satisfactory.

Such an analysis often produces remarkably useful surprises. Several studies of service businesses have found that one factor strongly correlated with customer satisfaction is whether or not an employee addresses the customer by name. That's why many successful banks, hotels, and airlines now insist that their people use the customer's name in every transaction.

Ultimately, whether you obtain your quantitative data simply by asking questions of customers or by sophisticated statistical analysis, and whether the results surprise you or not, measuring customers' real desires is the only legitimate foundation for any measurement program. It's an essential step in focusing on the customer.

■ Customers' Desires Establish Standards

Once you've learned what customers want, you can gain a competitive advantage by making the customer's desires into **standards** for the performance of your business—in every detail, from your way of greeting customers to your way of sending bills.

Without a standard, customer satisfaction is left to chance. A standard, on the other hand, creates a clear target for your people. Suppose a supermarket's research shows that customers become dissatisfied when checking out takes more than six minutes. Then a standard that says:

Check out in six minutes or less for at least 98 percent of customers

will help the supermarket to consistently deliver what the customer wants.

As discussed in the previous chapter, each standard should meet the criteria summarized in the acronym MARC: Measurable, Achievable, Relevant to customers, and Controllable.

And in order to achieve standards that fulfill customers' desires, you'll also have to establish **process performance standards** for the people who provide the inputs for those serving the customer. Each process standard must meet the MARC criteria too. Measuring and controlling process performance can be just as important as measuring the organization's performance for its ultimate customers.

For more subjective quality characteristics, an appropriate standard might be "95 percent of customers rating performance 5.2 or better on a 1-to-7 scale."

When you set standards, **focus your organization's effort on the issues that your research has found to be most important to customers.** If you've identified fifty product and service characteristics, make yourself a world leader on the five that matter most to your customers. After you've done that, make yourself the leader on the next five.

■ Customers' "Surrogates" for Quality

In setting standards, bear in mind not only customers' desires but also the "surrogates" for quality that customers use in making their buying decisions. Customers often don't know enough about products or services to make truly informed decisions about their quality. They therefore look for factors in their relationship that *they* can judge. These may include:

In judging a lawyer . . .	how soon he or she returns telephone calls.
In judging a doctor . . .	how well he or she explains the procedures that will be followed.

In judging a car . . .	sound of the door slamming or appearance of the finish on the dashboard.
In judging a packaged consumer good . . .	absence of artificial ingredients.

Merely achieving high quality on the surrogates won't win you success. If you don't also deliver top quality on the other aspects of the product that customers care about, they'll soon leave you. But understanding how customers assess your product is crucial to creating customers.

■ Review Quality Characteristics at Least Once a Year

Never let either the product and service characteristics you track or the standards you set become cast in stone. Review them in discussions with customers at least once a year. Customers' tastes and preferences change, and so you too must change.

Monitoring and achieving quality characteristics will help you to exceed today's customer expectations. But you'll achieve the biggest growth in market share by meeting expectations and needs your customers don't yet know they have.

If you're ready to recognize new wants and change your performance accordingly, you have a big head start on your competitors.

■ ARE YOU DELIVERING WHAT YOU KNOW THE CUSTOMER WANTS?

Once you've established standards, you need a continuous system of measurement.

You can measure the achievement of some product and service characteristics directly, and for others you can learn whether you've been suc-

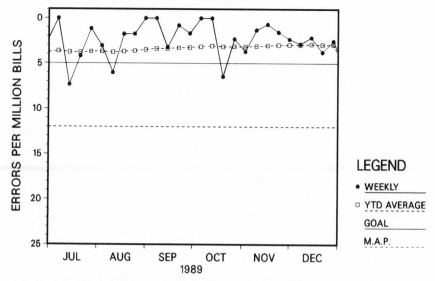

Figure 1   Currency Error Ratio per Million Bills Processed

cessful by watching customers' behavior. If a bank is measuring the service characteristic, "errors on bills processed," for instance, it will simply count the errors reported by customers. See Fig. 1, prepared by First National Bank of Chicago on the accuracy of its currency processing in the last half of 1989. The bank was making about 2.3 errors per *million* bills when "minimum acceptable performance"—here, the performance the bank believes is typical of other banks—was 12 per million. It had already exceeded its interim goal of 5 errors per million.

Joban Spa Hawaiian, on the other hand, judges the quality of its food by tracking how much customers eat. If they eat at least three-fourths of the food on their plates, Joban assumes they liked it.

But you won't discover how you're doing on the majority of characteristics that mean most to customers unless you develop a reliable method for learning customers' opinions. Four of the commonest methods are:

1. Customer comment cards provided when the product is delivered.
2. Mail surveys of customers.
3. Personal interviews—face to face, by phone, or in a focus group.
4. "Mystery shoppers."

Measure, Measure, Measure   165

Comment Cards

Customer comment cards are the easiest way of measuring satisfaction, and they can provide a direct link with every customer. They give you unfiltered feedback that will warn you about problems demanding quick action.

But comment cards are a passive way of communicating. You're doing nothing to encourage the average consumer to fill in the card, and so you'll hear only from customers who are motivated to respond. Certainly, they may not be typical.

If you're using the data from comment cards to determine pay and promotions, watch out. Managers can easily influence the results, even "stuffing the ballot box" with favorable comments.

You can't avoid the bias just by providing a postpaid mailer and asking customers to mail the form in. At one hotel chain, managers have been known to try to affect the comment-card results by standing at the checkout desk and solicitously asking departing guests whether they have enjoyed their stay. If the guest responds with a complaint, the manager will commiserate and promise to try to correct the problem. But if the guest responds with praise, the manager will present a postpaid comment form, saying, "Thank you very much. We'd appreciate it if you'd fill this out and mail it in."

Despite the cards' limitations, if everyone in your organization sincerely desires to learn what customers think, and your budget is limited, comment cards can be an effective way to learn the customer's desires. Moreover, some successful companies use comment cards *in addition* to more scientific methods. Create a *quick* feedback system that will fix problems reported in comment cards so fast that customers will see their comments have an effect. That will make customers loyal. And comment cards are an extremely inexpensive way to measure customers' satisfaction.

Mail Surveys

Surveys by mail often cost only a bit more than a well-managed program of comment cards. Regular customers will usually respond to a mail survey. A data-processing manager for a company that uses your computer time-sharing service will probably appreciate being asked for opinions about your service and will readily return your form. Customers who feel less tied to you—say, people who spend less than $50 on your product—may respond less dependably. But you may be able to increase response rates either by including a "bribe" such as a discount on the customer's next order, participation in a drawing for a gift, or by making follow-up telephone calls to those who don't reply.

Mail surveys do have flaws, however. Many of them get answers from only a small portion of the people solicited. Although the results of a mail survey probably are less biased than those of a collection of comment cards, it can be hard to know how representative your sample is.

Customer Interviews

If you employ interviewers, you'll have more control over who responds. Customers are generally more likely to answer questions posed by a good interviewer, either on the phone or in person, than a survey sent by mail. Interviews can frequently give you quicker results than comment cards or mail surveys. You can often launch an interview campaign on a Monday and have the results by Friday.

Moreover, interviews allow you to "branch" more easily than other survey methods. Suppose you want to ask a specific question of all customers over age sixty-five who heard about your product through a friend's recommendation. In a mail survey you can ask:

> If you are over sixty-five years of age and you learned about our product from a friend, please turn to page 6 for additional questions.

But if you include many such details, your questionnaire is likely to read like an Internal Revenue Service form. A common response is: Forget it. They won't fill out the questionnaire. On the other hand, if you hire interviewers to talk to your customers, it's fairly easy to train them, say, to ask the questions on page 6 if they receive one set of responses to their original questions, or the questions on page 7 if they get other answers, and the questions on page 8 if they receive a third.

Of course, interviewers are human. Their biases may differ from those of your customers, and sometimes will affect the answers you get. Also, interviewers may take shortcuts to complete their work sooner.

Mail surveys and interviews are valuable sources of customer-keeping information. But their quality has to be monitored with care.

Mystery Shoppers

"Mystery shoppers" are professionals—unknown, of course, to the people being evaluated—who pose as customers and then report on service. They are often quite reliable. Let's say that the same five mystery shoppers visit a bank branch in Forrest City, Arkansas, and another in Little Rock, 100 miles away. If all report that the service in Forrest City was better, then you have solid reason for believing it really *was* better.

But mystery-shopper programs involve four potential problems:

1. The cost is **much higher per encounter than for other forms of measurement.** An interviewer can visit a supermarket and interview dozens of customers in one day. But it's hard for a mystery shopper to do business with more than two or three employees on any visit. Thus many companies judge employees by such a mere handful of mystery-shopper encounters, far too few to constitute a statistically valid sample. If a clerk is a dynamite worker eight hours a day and the mystery shopper happens to arrive just after a bout with the flu or a hassle with a ridiculous customer (there are such things), the company often gets a report that's much more negative than the clerk's real performance justifies.

2. **Employees may catch on and figure out who the mystery shoppers are.** Sharp employees may give the mystery shoppers special treatment.

3. **Mystery shoppers themselves may be inconsistent or biased.** Reports from the same mystery shopper at different times, and from different

mystery shoppers in similar situations, both need to be carefully analyzed to determine how consistent they are. Mystery-shopper reports need to be compared with actual customers' evaluations to determine whether they may differ.

4. **Even if the mystery-shopper system is fair, employees may feel that their privacy is being invaded.** Employees who find out they've been visited by mystery shoppers can sometimes feel manipulated—and resent it. Big Brother again.

Choose the method of measurement (or combination of them) best for your needs. Then closely monitor the quality of your research.

Other Measures

In addition to creating your own measures of customer satisfaction, carefully track any measures that are independent of your own organization. Pay attention to:

- industry-wide surveys such as the J. D. Powers reports in the auto industry,
- customers' supplier-evaluation systems,
- surveys of dealers and suppliers,
- the media, and
- government agencies

And know what they're saying about your product and about your competitors.

■ Should You Create Customer-Satisfaction Indexes?

Some organizations can focus most precisely on the customer by creating a Customer Satisfaction Index (CSI) number that provides your best estimate of how well you're satisfying a group of customers. Some manufacturers in the auto industry have created a CSI to track each dealer's performance in servicing customers, for example. It summarizes many measures that the automakers collect, such as the number of complaints the parent company receives, satisfaction with the dealer as reported in surveys, number of problems found in inspections of the dealership, and proportion of vehicles repaired by the dealer that later have to be repaired again.

People can't keep track in their heads of all the data indicating how well an organization is serving customers. An index number then can be an excellent tool for focusing people on the customer's most important needs.

Creating a CSI forces the automakers to think through customer-satisfaction issues carefully. The index helps them compare dealers fairly and provides an exceptionally clear warning of customer problems. A single number that indicates customer satisfaction, moreover, gives that aspect of service a "hardness" that helps people take it as seriously as the current month's sales and profits.

Begin creating Customer Satisfaction Indexes by analyzing how many kinds of customers you have. Sea Land Corporation, a New Jersey-based container freight company, concluded that it had two crucial types of customers: its shippers and the people receiving the goods. Moreover, each group valued different aspects of service. The shippers cared about prompt pickup and error-free billing, and the recipients cared about predictable, courteous delivery. The company thus created a different index for each group.

A CSI is no panacea. It benefits a company as long as it doesn't become an end instead of the means to a happier customer. Every index needs to be reexamined at least once a year. Like any other number, the CSI can be manipulated—some auto dealers have been known to send a bouquet of flowers to customers in the week before the manufacturer is to conduct a survey it will use to calculate the index. Give your index especially careful scrutiny every time your company faces a hard-to-understand problem such as loss in market share.

Used correctly, a CSI or a similar number can truly help your organization. At a well-run Marriott Hotel I was served by a waitress wearing a button that said, "What's my GSI?"

I asked her: "What *is* your GSI?"

She replied: "It stands for 'Guest Satisfaction Index.'"

"And what is yours?" I asked.

"Our GSI is 96.5," she replied, with obvious pride. That index helped her to feel pride exactly where she should have felt it: in satisfying customers.

I asked, "Do you do anything differently because of it?"

She said, "Well, I guess I do a better job for you."

■ TAKE CUSTOMER-FOCUSED MEASUREMENT TO PROCEDURES EVERYWHERE IN THE ORGANIZATION

Measuring the customer's experience alone isn't enough. Effective companies also link measurement of internal processes in every part of the organization with customers' expressed needs. They track the speed with which airplane cleaners take trash off planes, and how long it takes to cook a hamburger. When they measure scrap rates in manufacturing, their people remember that the reason is not just to save money, but also to create an excellent manufacturing process that will give customers what they want.

At Joban Spa Resort Hawaiian, the guidelines for the presidential "diagnosis" of every department help determine whether the department is:

- measuring and controlling the right things,
- charting them appropriately,
- setting appropriate goals for the entity that is measured, and
- setting appropriate "action limits"—amounts of variation from norms permitted before management acts.

Most companies do too little measurement at all levels, and waste much of the measurement effort they do make. Why? Because they decide what to measure mainly by whim or by staying with "what we've always measured." In short, they miss opportunities to use measurement to benefit the customer—to thoroughly track the procedure of fulfilling customers' desires.

Measure, Measure, Measure   171

Yet each group within an organization can improve by continuously measuring the procedures that matter to both internal and external customers. First National Bank of Chicago's measurement program began early in the 1980s when it was struggling to recover from poor profits and a difficult transition to new top management. After a dispute about what should be measured, Larry Russell, who headed all operations of the bank except lending, declared: "No one in this room knows. The only people who know are customers. Go out and ask them."

They did. And First Chicago used their customers' comments to develop some 500 quality measures throughout the non-lending side of the bank—measures such as accuracy in automated clearing-house transactions (in percentages) timeliness in corporate trust transfers (percentage within seventy-two hours), and an index to quality of shareholder's communications (calculated with a point scale that measures accuracy, completeness, and courtesy). Every group in the bank tracks its own performance with charts such as those on page 165. The result: A customer-focused quality program that penetrates to every corner of the corporation.

When you've determined what customers want you to do, use that knowledge to determine what should be measured and charted within your organization and workgroup. That decision is crucial to implementing any customer-driven vision.

■ TELL THE WHOLE COMPANY THE NEWS

Once you've got the data, use them.

Departments need every scrap of raw data relevant to their own parts of the business. Teach people to plot graphs and charts of their progress and their failures. "It's exciting to go into a plant," says Joel Ramich, associate director of quality at Corning Works, "and have a guy who's maybe been there twenty years and never said anything come up to you and say, 'Mr. Ramich, would you come over here and look at my charts. I want to show you something.'"

Some corporate publications, traditionally filled only with "good news," make an excellent forum for telling both good *and* bad news from customer surveys. The publication of British Airways' vast Ground Operations-London group, *GOAL*, carried a chart showing the exact causes behind delays suffered by passengers. It didn't attempt to assign blame—it just reported the facts. But making the facts widely available showed that management was serious about attacking the problems, and encouraged the groups causing the largest slice of the problem to attack it with extra vigor. Also, *GOAL* carries extensive reporting on actions of British Airways' competitors.

If you don't have a glossy publication, however, a black-and-white report produced on a computer printer will communicate what customers are saying just fine. Most people hunger for trustworthy information about their company. If company leaders cite the statistics in the report and make clear that they're using them in making decisions, others throughout the organization will believe they're worth attention.

■ Deliver Data—Whether Wanted or Not

Successful leaders insist that their people take customer data seriously. One of the most interesting assignments I ever had came from General Motors of Canada when it set out to deal with a group of dealers who had poor Customer Satisfaction Index ratings.

Ordinarily, automakers have remarkably little leverage over dealers who neglect customers and show no interest in improving. Short of refusing to renew a franchise—a move that would anger many good dealers as well as scare bad ones, automakers believe there's almost nothing they can do.

But GM decided this impasse was wrong. It told thirteen low-customer-satisfaction dealers that if their ratings didn't improve, their franchise contracts would be in jeopardy. And to provide motivation and help, it created a special five-day seminar to be held at Disney World in

Florida, where they'd be forced to confront the implications of their poor performance. Attendance by the dealerships' owners was required.

This wasn't any junket. Before the trip, the dealers had to either sit in or listen to focus groups in which their customers discussed the anguish they'd felt trying to cope with botched repairs and unresponsive dealership people. Customers' problems started to become real to them. At Disney World, they spent their time listening to careful reports on the tough economic future of the auto industry and about how to make service a competitive weapon. I spoke about the successes and strategies of today's customer-focused companies. Disney's staff talked about service-management skills.

At the end, we saw an obvious change in attitudes. Most of the dealers truly seemed to believe in excellent service. Said one:

> I'm at the time of my life where I don't like being told what to do. I didn't like being told that I had to come down here, and when we started Sunday night, I did not want to be here. But the last five days have been very significant for me. And you were right in forcing the issue.

The concrete results of the program so far are impressive. Eleven of the thirteen dealers show major improvements in customer satisfaction.

■ Use Measurement to Change Behavior

General Motors of Canada is an example of how good measurement, effectively communicated and used, changes people's behavior. It helps them serve customers better.

Perhaps a more typical example is First Chicago Corporation's experience. It found its measures helped its managers manage at every level and also showed people that the bank was serious about improving.

Managers established a meeting every Thursday afternoon to discuss progress, and they actually opened it to the bank's customers. The customers could see that something new and different was happening. "It's real hard for people to believe that our senior management sits there from 1:30 to 4:00 every Thursday," says Aleta Holub, First Chicago's manager of quality assurance.

First Chicago is careful how it uses its numbers to manage quality. "No one ever gets shot for having a problem," Holub adds. "The only thing you would ever get shot for is not having an action plan in place. That's important because if you shoot people on the basis of quality numbers, they'll become very creative in finding ways to play with those numbers."

But today serious efforts have produced dramatic improvements. The bank has been turned around. Whereas in the early 1980s return on equity hovered around 7 percent, by 1988 it rose as high as 24.8 percent. Recent write-offs were smaller than those experienced by many other money-center banks, and earnings from corporate non-credit, the part of the business most affected by the quality program, are significantly above those of the past and of the competition.

Most important, First Chicago's customers are happy. Today, First Chicago has turned its data into a powerful competitive weapon. Salespeople take a "chart book" when they call on a potential customer. It shows graphs of First Chicago's performance on sixty measures of performance for the customer. It's a powerful tool in obtaining new business. First Chicago's customer-focused measurement has made a difference.

First Chicago's experience, though, is rare. Too many companies remain slow to measure what's happening in their corporate lives and communicate it to employees and customers. Some managers fear what the press may say if they report bad news to their employees and reporters find out. Others fear looking bad in front of their own people. And many companies are so accustomed to top-down decision making—leaders telling people what to do rather than *asking* everyone in the organization what should be done—that it doesn't occur to them that people in the organization could make good use of more and better data.

But everyone needs data. Good data, properly distributed, transform organizations.

- Begin a comprehensive list of the product and service quality characteristics that your customers—either your organization's final customers or the internal customers of your workgroup—are likely to seek in your product. Or begin a careful review of your current list. Communicate with customers in open ways such as interviews and visits to customers, to encourage them to add characteristics to the list.
- Once a list has been prepared, survey your customers to determine the level of performance that is satisfactory for each characteristic and the relative significance of each characteristic.
- Establish standards for your organization based on what customers have told you is important to them. Become the industry leader on the top five characteristics.
- Establish a reliable, continuous measurement system to determine whether you're achieving your standards.
- Review how you measure processes inside your organization. Establish new measurement programs driven by your need to achieve what you now know the customer wants.
- Tell everyone what your measurement system discovers—both good news and bad.

■ RESOURCES

Arlene Fink and Jacqueline Kosecoff, *How to Conduct Surveys: A Step-by-Step Guide* (Newbury Park, Calif.: Sage, 1985).

William H. Fonvielle, *From Manager to Innovator: Using Information to Become an Idea Entrepreneur* (Trevose, Penn.: AMS Foundation, 1989).

Floyd J. Fowler Jr., *Survey Research Methods*, rev. ed., Applied Research Methods Series, Vol. 1 (Newbury Park, Calif.: Sage, 1988).

Robert W. Hull, H. Thomas Johnson, and Peter B. Terney, *Measuring Up* (Homewood, Ill.: Richard D. Irwin, 1991).

Richard M. Jaeger, *Statistics: A Spectator Sport* (Newbury Park, Calif.: Sage, 1988).

Paul J. Lavrakas, *Telephone Survey Methods: Sampling, Selection, and Supervision*, Applied Social Science Research Methods, Vol. 7 (Newbury Park, Calif.: Sage, 1987).

■ NOTE

1. Milind M. Lele, *The Customer Is Key*, (New York: John Wiley, 1987), p. 111.

*Lech Walesa told Congress that there is a
declining world market for words. He's right.
The only thing the world believes anymore is
behavior, because we all see it instantaneously.
None of us may preach anymore. We must behave.*
—Max Du Pre, Chairman,
Herman Miller[1]

7

Walk the Talk

Art Wegner learned to be a leader at age twenty-two when he became an ensign in the U.S. Navy. He was sent to a nuclear submarine and given command of the torpedo room.

"I'd hardly seen a torpedo," he says. "And I'm supposed to be in charge?"

Yet Wegner quickly found that he could indeed lead the group. The chief petty officer, a wizened veteran of twenty-seven years in the Navy, ran the nuts and bolts in the torpedo room. Wegner promoted his commander's aims, encouraged people to improve their skills, celebrated team successes, and dealt with sailors who would test the system by "forgetting" to follow orders. He showed his interest by his constant desire to learn and to listen to his people's problems.

And he set an example of principled behavior for his men. One man had just come from a "Retraining Command" run by the Marine Corps, where he'd been sent because of a vicious temper. The problem was still with him. "Literally, I saw him take a great big monkey wrench like he was going to throw it, and instead of throwing it you could see him grab control of himself, and he took the handle and actually bent it," Wegner recalls. The twenty-two-year-old Naval Academy graduate talked with the man about his experience in Retraining Command in an open, honest way. As they talked, Wegner could see the man coming to grips with the temper problem. His behavior improved.

Wegner could succeed only by *embodying* the Navy's goals. "You were put into a situation where you had to get your job done through others' efforts. That's where a lot of people come unstuck in big business," Wegner says today. "They may be very smart, but one smart person is not as effective as ten or a hundred people all going in the same direction."

Two and a half decades later, Wegner was named president of Pratt & Whitney, the jet-engine manufacturer. Pratt engines powered nearly 65 percent of the world's commercial jetliners. But the company appeared to have lost its edge and badly needed a sense of purpose and direction.

Over the years, Wegner had learned that leading the 45,000 people of Pratt & Whitney required essentially the same kind of efforts that he had made as a twenty-two-year-old leader of a gang of seamen. Wegner had to *walk* as most business leaders *talk*—to put the customer first; to promote constantly his vision of the organization, to be an example of the kind of life that embodied the organization's goals.

One of Pratt's symptoms when Wegner became president was that the company handled customer problems with indifference.

Over the years Pratt, largest division of the conglomerate United Technologies, had displayed arrogance. Technical expertise had taken on more weight in promotions than ability and willingness to show people what needed to be done for the customer. As a result, a customer calling with a problem might wind up talking to a talented but insensitive engineer who would respond to complaints with a comment like, "You don't know what you're talking about." Simple spare parts such as hoses might take six months to reach a customer.

Surprise: customers were angry. The United States military took Pratt's $1 billion-a-year contract for the engines that powered the F-15 and the F-16—fully one-fourth of Pratt's revenues—opened it to competition, and gave General Electric's jet engine division 75 percent of the business in the first year of the new regime.

Commercial airline customers were upset, too. Pratt had beaten General Electric continuously since the 1960s, but in their anger customers welcomed new GE engine designs gladly, and GE was soon selling far more engines than Pratt.[2]

Against this background, Wegner's first years at the helm were painful. Congress summoned him to a public hearing to explain why Pratt was charging the Navy $18 each for replacement bolts. "The Navy had placed a special order for about eighty bolts of a type that we hadn't made in years," recalls Wegner. "I finally got through to them when I picked up a copy of the *Washington Post* and said, 'Look, this costs you fifty cents. What do you think it would cost if you had it special-ordered and they printed only eighty copies just for you?' "

Still, Wegner acknowledges that although the Navy created that problem, plenty was wrong with Pratt. He not only had to spend his time making changes, he also had to apologize in person to dozens of customers.

In the midst of all this turmoil, Wegner also found time to lead. As president of Pratt's manufacturing division before he became company president, Wegner already had begun the transition to a well-led organization by helping the manufacturing and engineering staffs to talk to each other. He worked on a program in which design engineers spent six months in the factory as supervisors, gaining experience much like his own as a rookie submariner.

"You can ask anyone who's been through it," says Wegner. "That six months is one of the best experiences they've ever had."

As president, Wegner made both personal leadership skills and technical excellence the bases for promotion and authority. And by constant repetition and his own example, he helped the company to focus on its customers.

Pratt launched an intensive, thorough training program throughout the organization to help people do a better job for all its people's customers, both internal and external. Outside customers came to Pratt's classrooms to air their gripes. Then the assembled Pratt people had to huddle on the spot and decide how to solve the problems the customers had just presented.

Pratt even ran ads in trade publications apologizing for its substandard service. One showed a jetliner tied down by a red tape, and declared: "Excuses just won't fly. Our bureaucracy shouldn't be your hang-up."[3]

Wegner decentralized decision making. He says:

> The most fun thing to do in the business world is to solve problems and make decisions. That's what a leader *wants* to spend his time doing. But what the effective leader must do is to avoid that and say to people who bring him an issue: "That's an interesting problem, and it relates to our key goals in the following way. Now let me know how *you're* going to deal with it."

That's how Wegner acted when he was a young ensign. The same approach worked just as well in mobilizing Pratt & Whitney.

"If I try to make a lot of decisions with the goal of reducing costs by 30 percent, I'm not likely to understand all the issues very well," says Wegner. "But if you get everybody—all those people in the organization—asking themselves, 'How am *I* going to get 30 percent of the costs out of there?'— the power of that is unbelievable."

Pratt has halted its decline and achieved a major turnaround. Today it equals GE's share of the market, and profits have risen dramatically.

Wegner and other United Technologies executives have shown what true leadership is and what it can do for any company. Sure, says Wegner, Pratt remains far from its old dominance, and its fortunes could decline again. But the changes at Pratt are real and dramatic.

And the lessons apply everywhere. The Forum Corporation recently studied the actions of leaders in twenty-three companies that, like Pratt, have taken giant strides toward becoming Customer-Driven Corporations. The sample included organizations with some of the most successful quality programs in North America, including three Malcolm Baldrige Award winners and a winner of Canada's Gold Award for Quality. We conducted structured interviews with leaders in each organization to learn which practices distinguished their leaders and made them effective.[4]

The results showed that the Pratt & Whitney experience is not unique. Seven leadership behaviors seem to distinguish the pivotal people in the successful organizations:

1. They **personally put the customer first.** They profoundly believe in the core values of customer-focused quality. They spend time with customers and become the voice of the customer in their own organizations.

2. They **promote their organization's vision,** not just by talking about it in ritual speeches, but by selling it, stumping for it, cajoling people in support of it, becoming evangelists for it, and setting up the organization's systems to deliver what the vision promises for customers.

3. They become **"students for life,"** constantly acknowledging what they don't know and seeking new ways of learning.

4. They **believe in and invest in their people,** training them, educating them, preparing them to do more than they can do today, and helping these people to use what they've learned to remake their own jobs.

5. They **make teams work,** first by bringing teams together from different parts of the organization to solve problems and then by giving their members the training they need, celebrating their successes, using them as communications channels between parts of the organization, and encouraging them to make decisions that will benefit the customer.

6. They **stay the course,** realizing that achieving customer-focused quality takes time and continuously encourage of others even when results have yet to appear.

7. They **live the organization's purpose,** leading by example, being willing to get their hands dirty, and continually creating more leaders within the company.

People who behave in this way can lead anything from a two-person work group to the largest businesses in the world. In this chapter we'll look at each of these behaviors of leadership in detail.

■ THEY PERSONALLY PUT THE CUSTOMER FIRST

The title on the business card of Anthony Harnett, chief executive of Bread and Circus, a Cambridge, Massachusetts, based "whole food store" chain, is:

CEO and Customer Service Representative

No false humility here. Harnett actually spends time at a vegetable stand serving customers. This part of his job, he feels, shows both employees and customers what's really important in his company.

Harnett is dedicated to an idea: "Food is life," he says, "and life should be about passion." He believes deeply in doing right by his customers. By spending time with them, he gets to know their own passions and see how they respond to his, in a way that no surveys could ever show. Moreover, when he spends time helping customers, Harnett also demonstrates his passion to the people who work for him.

Forum research found, to no one's surprise, that the companies making real progress in serving their customers are led by people who themselves put the customer first. More than that, they believe passionately in giving the customer what he or she wants. They spend much of their time with their customers and they become the voice of the customer within their organizations.

The successful leaders, we've found, are people who've made the move that one top executive called a "leap of faith": They didn't calculate how much quality would improve profits and then set out to improve quality because of those calculations; not at all. These people came to the profound, intuitive recognition that profitability is simply a lagging indicator of how well you're treating your customer. If they could create happy customers, then profit would ultimately take care of itself.

Westinghouse Chairman John C. Marous expresses the passion of the successful leader when he says:

> Total quality is everything, it's everybody. And it *is* almost like a religion. To have total quality you're going to have to change your culture. You can't change your culture without an emotional experience.[5]

Becoming the voice of the customer in an organization can be the most time-consuming part of a successful leader's job. William Siefkin, manager of the sales-development program at E. I. du Pont de Nemours & Company, said this when we interviewed him:

> The biggest challenge is to really bring the voice of the customer into your company, so that the customer is at every meeting in some way and so that our guidance comes from them when we're trying to make the decisions.

If there's a question about spending money, for instance, the best leaders always ask: How can we deal with this issue in a way that will serve our customers in the long run?

At Stew Leonard's in Norwalk, Connecticut, one of the best-run, highest-volume food stores in the world, we talked to Stew Leonard, Jr., son of the founder, who is now running the place. Like Harnett, he spends most of his time on the floor of the store. "All the other decisions distort your thinking," he says. "You know, like cost and volume and returns and all that analysis. If you do what the customer is requesting, all those financial results will follow."

That's what all the successful leaders we interviewed believe.

■ THEY PROMOTE THE ORGANIZATION'S VISION

"Good business leaders," says Jack Welch, president of General Electric, "create a vision, articulate the vision, passionately own the vision, and relentlessly drive it to completion."[6]

Successful leaders never stop selling their vision, goals, and strategy to their people. They're not afraid to promote their vision—almost like, say, Procter & Gamble promotes toothpaste. Says Pratt & Whitney's Art Wegner:

> You have to sit down and define a few attitudes and goals you want for the organization. Then you have to communicate them forever. Use every communications vehicle that you can. It's got to be repetitive—over and over and over. It's very time-consuming, but it's the only way to get where you want to be. Eventually people say: "This guy really means what he says. And he believes in it and he knows what he's talking about."

Effective leaders repeat themselves—not because their people lack the intelligence to understand the first time, but because repetition is necessary to communicate the full power of vision. It demonstrates seriousness.

"If I have any message," says Wegner, "it's this business of being persistent. The people in our companies are good people. The biggest problem in United States companies is management. We're not secure in ourselves and so we're afraid of regularly standing up and saying, 'I need help.' "

Chapter 1 of this book discussed the power that a customer-keeping vision can give an organization. A true leader promotes the organization's vision by constantly keeping it on employees' minds. Remember the signs that IBM's creator, Thomas J. Watson, Sr., used to promote his vision of a constantly improving organization: THINK.

The most successful top managers establish high-profile programs to teach an organization's vision, its basic beliefs, team-oriented work methods, and the ways in which the organization must change. They devote the largest portion of their time to keeping and promoting the vision. And they make radical structural changes to serve customers better.

They make sure people give the programs the time and resources they need to be successful. American Airlines conducted a campaign among its managers called "Committing to Leadership." The British Airways management crusade was called "Putting People First." Imperial Chemical Industries sponsored a drive called "Minding Our Own Business"—devoted to identifying what ICI's businesses really were and helping ICI's people to focus on them.

But signs and speeches, though important, aren't enough. The successful leaders don't just pay for the programs; they're active participants. Watch the leader of a company or a workgroup during a program that's designed to teach the group's people any of the principles in this book. You'll soon know a great deal about the future of the organization.

Recently we presented our Building Customer Focus workshop at Corning. Corning's chairman, Jamie Houghton, attended the first two-day course without interruption. He also has attended the first presentation of each of the major company-wide quality programs created by Corning's quality staff. "That did us a lot of good when we wanted to present them at the plant level and the plant manager said he was too busy to attend himself," says Joel Ramich, associate director of quality.

Many leaders don't understand why *they* need to attend a session that will teach a vision they already understand. One leader attended a course I taught—but brought with him a portable cellular telephone. While the program was in session, he would push back his chair and make calls on the spot, as the spirit moved him.

That kind of casual—and uncivil—attitude doesn't make sense. If they are discussing an issue you truly consider vital to your vision, doesn't it make sense to be listening intently? That's the only way to check whether the vision being imparted is truly yours. More important, listening to the presentations puts you in tune with your people. You'll share the principles, ideas, and anecdotes they're hearing. And, of course, by listening carefully you set a good example for the rest of your staff and demonstrate to *them* that you consider what's being said to be crucial.

Perhaps your most significant way of setting direction for your organization or your workgroup is your use of your time. Almost any decision and almost all information-gathering work can be delegated. But you can't delegate:

- the creation of a vision;
- the long, sometimes tedious hours spent understanding your people and explaining your vision to them;
- days spent meeting with customers to understand their problems.

Don't just *tell* people your vision, *show* them what you mean. A colleague of mine—John Humphrey, chairman of Forum—was named chairman of the Boston Ballet Company at a time when the organization was suffering a major deficit. Empty seats plagued its performances. To understand what was going on, he climbed to the top balcony to see what customers saw. He found upper-seat customers had to walk past dreary old scenery stored in the halls. Markings on the stage floor, designed to help dancers find just the right place to stand, were made with scruffy, distracting masking tape. They were an eyesore for anyone in the upper balcony even though they were invisible from lower-level seats.

It was apparent that the ballet company's staff had been designing productions with little consideration for the upper-balcony patrons. The hall had no refreshment facilities and, perhaps worst of all, at curtain calls the dancers bowed to everyone in the audience except the customers in the upper seats.

Humphrey pointed out the problem to the company's managers. They promised to fix it, but nothing happened. He told them again. Still nothing.

He then invited one of the managers to enjoy a performance with him—in the top balcony. The manager had to experience what the customer experienced, and he got the point. Within a week, the grubby old disused sets were gone and the distracting floor markings replaced with unobtrusive little crosses. Other changes were made, too. The focus was on the customer.

Has this shift in values made a difference? You bet it has. The Boston Ballet Company now enjoys a superior artistic reputation, sells out its performances, and runs a financial surplus. *Horizon* magazine cited it for "achieving one of the most phenomenal turnaround's of the decade in the art world."[7]

Promoting a vision is more than just selling. It also means demanding needed changes in outmoded, inefficient structures that have become comfortable for many employees. Sarah Nolan, president of Amex Life Assurance Company, recalls:

> We had not stripped away and looked at the essentials of this business in a very long time. So we put together a model office group and moved them out of the building. There were five of them, middle-management level. There was a marketing person, a customer-service person, a systems person, a new-business processing person. . . . I said to them, "All I want you to do is rethink the business of accident insurance. I want you to rethink the essentials. The only rules are: First, don't analyze anything we do in this building; you'll waste your time, you know it's outdated. Second, develop it from the customer's point of view. Third, don't use any systems we have in this building. And fourth, design the work as if you yourself would do it.

> They rethought this part of the business from the group up. They didn't draw an organization chart that was hierarchical. It was a circle. It had the customer in the organization chart. They created a team-oriented approach.

Anyone can become a leader, not just of the workgroup but sometimes of the organization as a whole. When Nissan Motors fell on hard times in the mid-1980s, groups of section chiefs—the lowest-level managers in the organization—established informal committees to promote closeness to the customer and creative thinking. The committees created a great variety of programs. They sent outstanding employees to the United States to bring the company closer to the customer. They allowed customers to get closer to them by letting the general public use the company's automobile test course on weekends. They published a "grass-roots" bulletin, in which they criticized their company's often-bureaucratic ways. President Yutaka Kume credited them with a major role in Nissan's turnaround later in the decade.[8]

■ THEY'RE "STUDENTS FOR LIFE"

I frequently ask people to think of the best leaders they've personally known, and then rate them (on a scale of 1 to 10) for charisma.

The average charisma rating for the best leaders is around 6. In other words, the best leaders are no more than average on the charisma scale.

How can that be? Isn't charisma—that extraordinary magnetism—a major element in leadership?

Apparently not. In fact, leadership actually depends on something quite different, which charismatic people often lack: an openness to new ideas and indeed an insatiable hunger for ideas that will help achieve goals.

The people who are successfully leading the transformation of corporations, we found, are "students for life." They're fun to be with. Because when they hear about something that is an aberration, rather than being threatened by it they go and learn from it. They say, "Here's an opportunity."

This sort of person is happy to spend time learning from customers, is willing to invest in getting customers to complain. When you interview the best leaders, they often get a little fidgety. They're so curious that they don't like to talk about themselves, because when they're talking about themselves they're not learning anything.

One "student for life" early in the twentieth century was Alfred Sloan, the man who created the General Motors management system. He would refuse to make any important decision if he felt he hadn't heard all the meaningful insights that the members of his organization could bring forward. He once short-circuited an overly quick consensus on a major decision with this comment:

> Gentlemen, I take it we are all in complete agreement on the decision here. . . . Then I propose we postpone further discussion of this matter until our next meeting to give ourselves time to develop disagreement and perhaps gain some understanding of what the decision is all about.[9]

Sloan actually refused to accept a decision his advisers urged until he could learn the diverse ideas they could bring to it.

He wasn't behaving like a stereotypical "forceful leader." But when leaders express quick, forceful opinions instead of an unquenchable desire to learn, they risk losing the benefit of much of their coworkers' knowledge and ideas.

This is an area where we see some real progress in the United States today. Dr. Thomas Malone, chief operating officer of Milliken & Co., the textile manufacturer that won the Baldrige Award in 1989, has had 6,000 executives come to study his company. During my research interview with him, he said:

> You know, Dick, what's really impressive to me—and I see it as a good sign—is that in the last three or four months, we've had more CEOs coming with the executives than we've ever had before.

These CEOs are going out to learn. And that shows they're recognizing that they have insufficient knowledge to become world-class competitors. They're becoming students.

■ THEY BELIEVE IN AND
INVEST IN PEOPLE

"Commitment must go both ways," says Charles E. Horne, vice president of Amica Mutual Insurance Company, a firm that consistently achieves exceptional service ratings from its policyholders. "We want people to commit to us, but we are committed to them in the same way.

Just as the best leaders have made a leap of faith in believing that doing right by customers will in the long run cause profits to take care of themselves, so too they've typically decided to believe in and invest in their people. After demonstrating that faith, they've seen the results in better work and increased profitability. In speaking of the people at Milliken, Dr. Malone said the difficult part of his company's transformation was

> to really believe that our people understood how to improve the processes and what we were doing better than anybody else, to enpower them to develop the solutions and to implement the solutions; and to do exactly the same thing in relationships with our customers—to listen to them, and really jointly with them, develop solutions to problems.

I mentioned in Chapter 4 that most Western companies spend far less than Japanese companies do on educating and training their people—as little as one-fourth as much. We found in our research that the transforming leaders are changing that. They invest in training that is focused on supporting the vision and on providing the skills that will enable people to fulfill that vision. "If that education and skill building isn't focused consistently with the vision, then it's going to be money wasted," says Joseph R. Bransky, director of corporate quality and reliability at General Motors.

Dr. Michael C. Vanderzwan, executive director of quality management for the Merck Pharmaceutical Manufacturing Division, says:

> You get confidence in your staff by training them. That's the key responsibility of managers: to train the people who work for them.

And once you've given them all the training that is necessary, then you've got to have the confidence that they're going to do the job, and empower them to do it. . . . You can't be afraid that your subordinates are going to be better at the job than you are. And they should be. We have a president in one of our divisions who has an excellent set of management philosophies, and one of them is that you should have working for you several people who are better qualified than you.

Many successful leaders are insisting that managers themselves teach courses or parts of courses. David Kearns at Xerox not only increased Xerox's investment in education and training vastly, but also presented the training to his organization by "cascading" sequence. Xerox decided to teach people skills in leadership and quality—like those we've been talking about in this book and those we list in the toolkit on pages 219–293. Kearns and the people who reported directly to him took the course first. Then they spent time using the skills and techniques. And then they themselves—all these top executives of a $17 billion-a-year corporation—actually taught the techniques to the people who worked for them. They worked with trained facilitators, but the managers had to do the first part of the training themselves.

Then the next level of managers taught the techniques to the people who worked for *them*. That got people's attention. You couldn't very well treat training as something nonessential if the teacher of your training course was a senior manager and you knew you would have to teach it to your own people. Other companies such as Milliken and Motorola have employed similar approaches.

▪ THEY MAKE TEAMS WORK

Few techniques of management have caused more excitement in the past decade than the creation of teams to set direction, share information, and solve problems. Only a few years ago, books about management taught how to run a hierarchy, delegate jobs, and manage subordinates. Although these subjects still are important, our research shows that today's managers succeed by deploying and effectively using teams that possess considerable freedom in achieving excellence for customers.

Work groups in these organizations function as teams in which the ideas of low-ranking people frequently count as much as those of the more senior members. Quality Circles, Customer Action Teams, Quality Councils, and cross-functional groups are constantly struggling to solve problems for customers and, in doing so, helping people in each part of the organization to understand the needs of the rest of the company.

At the same time, one of the most challenging tasks facing a leader is to make those teams work—to make them effective. Too often, employees who've suffered through management efforts to create teams tell stories of boredom and frustration. Remember the oft-told bromide about a camel being a horse designed by a committee? Many meetings that are supposed to enlist people in a common purpose or to make important recommendations or decisions are too often viewed by their participants as a waste of time.

Why can some leaders create productive teams, while meetings in companies and work groups led by others are a source of dread for employees? A few years ago Forum conducted research addressing this basic issue: "How does a person in an organization achieve **influence**—effectiveness is getting things done for the customer while working with people over whom he or she may not have operating control?" We looked at people who were excellent with influence as well as those who were below average at influencing others. Our goal was to get a clear picture of what the superior influencers were doing that distinguished them in their organizations. We carried out our first research on this question more than ten years ago, and since then have administered questionnaires to more than 4,000 managers.[10]

By studying people whose peers reported they were influential in teams, we made striking discoveries:

- People who are successful influencers achieve this influence through a sequential process. First they *build* influence, becoming an efficient team leader or member. Only then do they *use* influence, working with a network to develop high-quality, creative decisions. And once they've become influential they employ specific practices and *sustain* their influence, creating a basis for consistent execution by demonstrating openness and gaining consensus.

- Influential people have specific underlying attitudes about how people work together. They appreciate the necessity of supporting and helping others, sharing power, and building trust. These attitudes are likely to be missing in organizations where teams function poorly.

The following self-test is based on the research. By understanding how you would handle the situations or relationships portrayed in the eight brief scenarios you can begin to get a reading on how effective you are likely to be as a leader or member of a team—and where you might improve.

Look at the test questions and make your decisions. Score it based on the answer key on page 207 at the end of this chapter. You'll be the only one who knows the results, so be honest with yourself. Choose your answers not based on the responses you suspect to be the "right" ones, but rather the answers which come closest to what you'd actually do or say.

■ The Influence Test (Answers appear on page 207)

1. You're a successful energetic, influential group leader. And the group you lead is a good one. But what would happen if you stepped away from your leadership role without warning? What would your group most likely do?

A. They would suffer from analysis paralysis and spend endless time clarifying procedures and rules.

B. They would probably take the initiative and work to establish workable, flexible ground rules early in the game.

C. They would go out of the starting gate at a fast, confident clip, but get lost somewhere along the way and find a need to constantly regroup in order to maintain proper direction.

2. You've got a problem. And the only way you're going to solve it is by getting the help and cooperation of your colleagues. How will they respond to your overtures for assistance? You know the answer.

But if the shoe were on the other foot, and we asked your colleagues how you respond to their cries for help, what would they say?

A. That I'm generally willing to go the extra mile to provide needed information—and provide it on time.

B. That I want to help but I'm so busy that I have little time to talk. And it takes my office more time than it should to supply the necessary information.

C. That I'm a person who has a multitude of responsibilities and is reluctant to take time out to help anyone else.

3. You are at a meeting of your peers where pandemonium reigns. It's the eleventh hour and after months of work, the problem still hasn't been solved. Accusations are flying, the situation is without order, and decisions must be made on the spot or the whole meeting will end in disaster. What do you do as a separate, but equal, member of the meeting?

A. You make a battlefield decision and take control of the meeting by restoring order and making the decision.

B. You suggest an agenda and get the consent of everyone present.

C. You participate by making helpful, informed decisions—without trying to alter the meetings's format.

4. You've worked for months getting your part of the project ready to present to the rest of the group. Just when you're ready to present your work, another group member asks you to make some major modifications to your information. The rework on your part will be enormous. What do you do?

A. Stall the other person and go to a higher authority for advice, direction, and possible mediation.

B. Request information about the change you're being asked to make and determine how the change will affect the overall project.

C. Call a meeting of the group and ask them whether last minute change—and request for changes—makes sense.

5. You're busy coordinating a meeting when, out of left field comes a suggestion that is badly timed, has little or no merit, and contradicts everything the group as a whole has just agreed upon. You hear gasps of disbelief from around the table. You feel everyone's eyes on you. How do you handle the situation?

A. You ignore the outburst and move the meeting along to another subject.

B. You firmly, but politely, look the speaker in the eye and say, "We've already been over this ground, George."

C. You throw the ball into the troublemaker's court and ask him or her to test their suggestion against the group's.

6. You've got to make a critical decision and make it fast. Sally's proposal is sound, straightforward, and while it's perhaps a bit conservative, it will definitely get the job done. Harold's proposal is off the wall. He's offered you a radical solution peppered with interesting ideas that, if fleshed out, just might make it all fly. A decision must be made. What do you decide?

A. Time is short. You go with Sally because she's a known quantity. Her idea is a good one. And besides, the last thing you need now is a personality conflict.

B. You push personalities aside and go with ideas and evidence instead. Harold just might have a better idea. But he'll have to work fast to make them work. And you'll have to work with him.

C. You give Harold's plan to Sally and ask her to incorporate some of his better ideas into her plan.

7. In the ideal influence environment, the influencer never feels like he or she is going it alone, never has trouble getting people to take risks and implement decisions, never has to solve all the problems single handedly because of not being able to persuade others to help. Which of the following choices best describes your own "influence" environment?

A. We're winning the battle—and the war. My people realize that it takes responsibility, risk, and extra work to make things happen.

B. We're surviving, but things could be better. The people around here have a thing or two to learn about teamwork.

C. Red tape still snarls up everything. And what gets done is rarely done right in the first place.

8. Your rise to the top didn't happen overnight. And all the things you learned in order to stay at the top—like influencing others to help you achieve your objectives—weren't learned in a day. From your expertise, what does a person have to have to be a master influencer?

A. Practice, observation, and sheer experience. And a strong belief in gaining and maintaining people's trust, leading without alienating, and sharing power.

B. It doesn't take anything special to influence. All you have to know is how to push the right people buttons.

C. You've got to be born with the ability to lead and influence. Most leaders had what it takes even before they entered the business world.

If you scored a perfect 80, congratulations. This indicates you work well in teams. You've gained people's trust. You can lead or be a member of a group without alienating others. You're able to get people to take risks. You're willing to share power and delegate responsibility.

Better-than average influencers scored between 50 and 79 points. You're pretty good at getting results from people over whom you have no operating control. But there are chinks in your influence armor. Perhaps you still have trouble gaining the cooperation of stubborn colleagues.

Perhaps you find it difficult to delegate to others some of those expendable responsibilities that are slowing you down.

Average influencers score 49 or less on the test. You're in the same boat as the majority of your peers. You're probably working by the old chain-of-command method: leader tells lieutenant who tells worker. You appear to have difficulty working effectively with those over whom you have no operating control. You still may need to be convinced that influence, when exercised with skill, can be a powerful tool.

If your score was below 80, take a moment and review your answers. What can you learn from the preferred or "correct" answers? What do the answers you selected tell you about both your strengths as an influencer and your opportunities to improve?

■ THEY STAY THE COURSE

Programs aiming at corporate transformation often follow the pattern illustrated in Figure 1.

When a program starts, it opens with a great deal of fanfare and excitement. People's expectations rise rapidly—and often continue rising. They're excited about doing a better job.

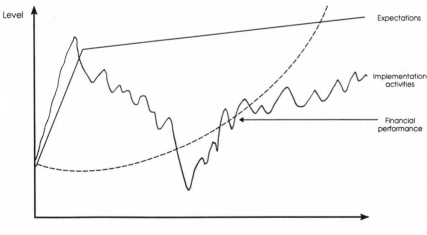

Figure 1 Stay the Course

But that doesn't translate to increased profits right away. Your program costs money, and it takes a while before many of your most important initiatives begin to affect the customer. Even if customers notice better service on the front lines immediately, they don't at once start spending more for your product.

After that initial push, moreover, the program has problems. A lot of work is being done, people are "turned on," but typically the company hasn't yet developed an integrated cross-functional effort, so that an effort by a group of people in the factory, for instance, may be defeated because some middle manager in the accounting or design department doesn't support it.

The result is that work seems excessive and people look for ways of reducing the effort they put into improvement activities. The whole program slows down. People feel frustrated and maybe cynical.

And in those hard times, most companies and most managers within companies back away from their programs. Leaders still talk about quality, but it becomes perfunctory. They start looking for an easier way to increase profits.

In these times, the leaders who will succeed distinguish themselves. Their leap of faith is real. If they're facing financial losses, they may make major cutbacks in areas that don't directly affect the customer. But their focus on becoming a Customer-Driven Corporation continues. They keep on pushing, keep on selling their vision to people, keep on celebrating the successes they've achieved and working to learn why those successes haven't influenced the customer as they should. The middle managers who will succeed keep fighting for the vision, struggling with others in the organization who've become cynical. Improvement activities in the organization gradually recover. And then, maybe as much as three years after the effort started, the effects show up on the bottom line.

It's like eating health food, or going on a long-term weight-loss program. You have to be persistent and you have to be patient. Tom Malone of Milliken said:

It takes two or three years for it to really happen. You've got to get the scoreboards in place, and then the environment of applauding people for improving the characteristics you're measuring on those scoreboards. You can document all that up-front cost. The payback is later.

You become a successful leader if, like Tom Malone and Roger Milliken, you persevere through the bad times.

■ THEY LIVE THEIR PURPOSE

All customer-driven companies are alike. Watch their leaders' behavior and you'll understand what they value and see why they succeed.

Sometimes living your company purpose can call for great sacrifices. Allan C. Emery, Jr., a prosperous Boston wool trader, closed his business in the 1950s because he saw that synthetics were replacing wool and low-cost foreign mills were replacing American spinners. Soon after, he agreed to open a joint venture to clean New England hospitals with a then-unknown midwestern cleaning contractor named ServiceMaster. He was to invest the largest part of his fortune in the project, which would be ServiceMaster's first venture outside the Midwest.

ServiceMaster's training methods for a new franchisee investing a fortune in a startup were consistent with the behavior it expected of its own executives, but they were unusual by any other standards. Emery recounted them later:

My first assignment was to work for two weeks as a houseman at a large metropolitan hospital. I was to mop corridors, empty trash containers, and clean ash trays. While not in the best condition for this work, I completed the day's schedule. The shock was not in the work but in the general rejection of me as a person because of my green uniform and the kind of work I was doing. Not a single person responded to my "good morning" except others in the

Housekeeping Department. I had never before experienced the caste system. . . .

As the days passed by, I learned a great deal about myself and the needs of people to be accepted. My work strung out into months of being a housekeeping aid, trash collector, linen distributor, linen sorter, floor refinisher, wall washer, and "unit check-out girl." During this period I made some wonderful friends. I began to understand Ken's [Ken Hansen, president of ServiceMaster] question to me earlier: "Allan, you are 45. You have time for one more career. Are you going to invest your life in things, commodities, or people?"[11]

ServiceMaster and its New England branch showed sales and profit increases of greater than 20 percent a year for decades, and the parent company eventually bought Emery out for a sum far greater than he had invested. By putting high-ranking leaders through months of training in green uniforms, ServiceMaster was forcing them to live its commitment to service. And it paid off: *Fortune* magazine has several times listed Service-Master as the most profitable large company in the United States.

"The man at the top of an organization must personally do things that others would hate to do most," said Soichiro Honda, founder of Honda Motors.[12] Frequently leaders don't have to do anything as time-consuming as Emery's months of green-collar work. But all leaders, whether they are company presidents or crew chiefs in fast-food restaurants, need to live a life that sets an example for their people. They have to *become* the kind of person the company needs to achieve its vision.

In our research, we constantly heard people saying just that about the best leaders: They *are* what they want their companies to *become*. When Ray Kroc built his first McDonald's restaurant, hundreds of other fast-food places already lined the roads outside most of America's major cities. But their owners were mostly looking for a quick profit. You wouldn't find them behind the counter or cleaning the bathrooms. How many of them are still in business today?

But Kroc's franchisees could never tell when Kroc himself would show up in their parking lot, picking up hamburger wrappers that the franchisee had neglected. He always lived his business vision—providing "Quality, Service, Cleanliness, Value" to ordinary people.

A businessman who's done the same more recently is Robert Galvin, who retired in 1990 at age sixty-seven as chairman of Motorola but who still heads the company's executive committee. In fact, Galvin has set an example not only for his own employees but for businesspeople everywhere. He spent the 1980s tripling the size of his company and in doing so created a powerhouse in two of the most competitive of all industries— semiconductors and mobile electronics.

Motorola began in 1928 when Galvin's father Paul set out to build radios that could be installed in cars. The name Motorola came from the first part of the phrase "motor car" and the last part of the word "Victrola," which was then synonymous with sound reproduction.[13]

Back in 1979, the company was number one by several measurements in every market where it competed. It was reporting record profits and excellent growth.

But at a problem-solving officers' meeting in that year, Galvin was challenged by Arthur Sundry, a Motorola communications executive. "Bob, we're talking about the wrong things," one executive recalls Sundry saying. "Our customers are telling us that our quality stinks."

It was a shocking statement. Perhaps the fundamental evidence of Galvin's lifetime of leadership is that Sundry dared to make such a declaration in a year when the company had good reason to be pleased with itself. At the time, Motorola had a reputation for superior quality. Relative to its competitors, at least, the reputation was justified. Yet by the late 1970s most United States companies could probably have heard the same message Sundry heard if they had listened to their customers. Paul Noakes, Motorola's vice president for external quality, recalls that Motorola prod-

■ Knockouts: The Signs That a Company Attempting to Transform Itself Is Likely to Fail

Many—perhaps most—of the businesses trying to transform themselves into customer-driven companies will fail. Over the past few years, I've discovered that a negative answer to any of the five questions below is a strong indicator of likely failure. Answer these for your organization:

1. **Is the person in charge of your customer-focused quality initiative widely respected in the organization as a person who can achieve results?** If he or she is a highly talented manager whom people trust, your effort stands a good chance. If he or she is an oldtimer awaiting retirement or a youngster with little credibility, you're in trouble.

2. **Is the CEO dedicating the time required to lead this effort?** If not, you've got to transform the CEO before you can transform the company.

3. **Do cross-functional teams work effectively?** Because many different groups must collaborate to deliver quality products and services, this is essential. But too often quality efforts are handcuffed by politics and interdepartmental competition.

4. **Is the company following its initial announcements and rallies with a systematic, integrated program for training and other transformation activities?** Kickoff sessions are important, but without proper follow-up they will ultimately leave an organization frustrated and cynical. You need a well-designed program that clarifies the organization's vision, develops and rewards people, transforms internal processes and measures, and promotes results.

5. **Does the company stick with the initiative when business slumps?** Many organizations on their way to great success hit crises. They may have to reduce some expenditures for customer-focused quality programs or spread their programs over a longer time. But the successful firms neither drop their programs to save money nor lose their commitment to them. Rather, they use their vision and customer awareness to turn the crises into opportunities to reinforce their objectives in people's minds.

ucts suffered from "infant mortality"—a distressing proportion failed soon after they were taken from their boxes. Moreover, Motorola often made annoying little process mistakes—building good products, then shipping them to the wrong addresses, or sending erroneous invoices.

Although most businesspeople probably would have given Sundry's assertion no more than a few minutes' attention, Galvin responded with a readiness to listen. When that became apparent, other managers added their own discomforts.

Galvin tore up the meeting's agenda and devoted hours to the quality problems.

By the start of the following year he had named a new director of quality—not someone from the technical specialty called "quality control," which at the time was a backwater that rarely attracted the best minds, but rather a top general-management executive. The new quality director, Jack Germain, had been assistant general manager of Motorola's communications sector, a $1 billion-a-year business.

Perhaps more significant, Galvin himself set out to visit a Motorola customer at least once a month. And when he did so, he didn't go to the customer's top managers. He went to people who used Motorola products. Then he would write memos for the rest of the company as detailed as a junior executive's trip report.

"I can tell you," says Noakes, "I was a general manager [in Motorola's communications business in Canada] and I was embarrassed—there were times he knew more about my business than I did. And so I realized there's a great learning process in visiting customers. I started doing the same thing. And now all the officers do the same thing, go out once a month and spend a day with the people who use our products."

Galvin isn't an engineer. He dropped out of college in the 1940s to join his father's business. But he also quickly realized that the techniques of process control were too important to delegate, and he studied them with the rest of his nonengineering staff.

He changed the agendas of Motorola's operating review meetings. He put quality and cycle time—the ability of each part of the organization to deliver what the customer needs quickly—*first* on the agendas of every business review instead of third after financial and operating issues.

Most dramatically, Galvin developed a new habit. After questioning his people closely on quality and cycle time throughout the initial segment of business reviews, Galvin would simply leave the meetings. He left the lower-ranking people to discuss the financial issues among themselves. What a message!

Today the "infant-mortality" problem is gone. By 1992 the company hopes to achieve what it calls "Six-Sigma Quality," which means that products will be 99.9997 percent error-free.

Motorola is an engineering-oriented company, and even today Motorola's quality of service may not quite equal its product quality. "Soft" aspects such as empathy of service personnel get less emphasis than reliability of hardware, for instance.

Yet even here, Galvin has been a leader. The practice of having executives visit lower-level customers opens the organization to the ordinary user's needs.

"He has us preaching almost like we're in a pulpit," says Noakes. "If we take care of our customers, if we have high quality and reduced cycle time and we satisfy our customers, we don't really need to worry about the business."

ACTION POINTS

- To become an effective leader, whatever your official title in your organization, start by determining what you value. Where does the customer rate in your own hierarchy of ideals? One way to determine what you value is to observe your own behavior. How do you spend your time? With whom? What really gets you upset? When business is tight, what do you cut? If someone were to follow you around and observe you for a month, what would they say you value? Then ask: Are your values consistent with your organization's vision? If not, they must be reconciled. Either change the vision or change what you value.

- Spend time with customers—an average of at least one day a week. Be the customer's representative within your organization at every meeting.
- Remember that selling your organization's vision to your people is critical to success. It is necessary and sometimes time-consuming. Repeat it over and over. See every event, every meeting, every communiqué as a vehicle for this important message.
- Evaluate how you spend your time. Can you delegate more information gathering and decision making? Your time is needed for communicating your vision, understanding your customers, and spending time with front-line people.
- Look at the organizational structures in your company or work group. What should be changed to serve the customer better? Not all time-honored structures work for customers. By changing them, you not only can create a better experience for customers but you also signal that nothing will be allowed to stand in the way of doing right by customers.
- Look at interactions with your people from the standpoint of what you can learn from each one, as well as what they need to hear from you.
- Be constantly aware that your people have basic knowledge that enables them to improve the processes of your organization better than anyone else. Free them to use this knowledge. Invest in them so that they'll have the skills to turn knowledge into results.
- Behave in a way that is consistent with the vision you have created. Let this behavior be clear and obvious so that those who see it can follow your example.

■ RESOURCES

Warren G. Bennis, *On Becoming a Leader* (Reading, Mass.: Addison-Wesley, 1989).

James MacGregor Burns, *Leadership* (New York: Harper & Row, 1978).

Max DePree, *Leadership Is an Art* (New York: Doubleday, 1989).

Joseph M. Juran, *Managerial Breakthrough: A New Concept of the Manager's Job* (New York: McGraw-Hill, 1984).

A. Zaleznik, "Managers and Leaders: Are They Different?" *Harvard Business Review*, 55, no. 3 (1977), pp. 67–78.

▪ NOTES

1. *Fortune,* March 26, 1990, p. 26.

2. A good summary of the problems and progress of United Technologies is Todd Vogel, "Where 1990s-Style Management Is Already Hard at Work," *Business Week* (October 23, 1989), pp. 92–100.

3. Vogel, pp. 92–100.

4. This research has also been summarized in "Leading the Customer-Focused Quality Company: Lessons Learned from Listening to the Voices of Leaders," Forum Corporation, forthcoming.

5. "Westinghouse Gets Respect at Last," *Fortune* (July 3, 1989), p. 98.

6. *Harvard Business Review,* September-October, 1989.

7. "U.S. Arts: Strategies for the 80s: Boston," *Horizon,* (November 1988); p. 41.

8. Yu Inaba, "Nissan's Management Revolution," *Tokyo Business Today* (October 1989), pp. 38–40.

9. J. Edward Russo and Paul J. H. Schoemaker, *Decision Traps* (New York: Doubleday, 1989), p. 145.

10. *Developing a Model of Influence Behavior: A Research Report,* Research and Development Group, The Forum Corporation, 1983.

11. Allan C. Emery, Jr., *Turtle on a Fencepost* (Waco, Texas: Word, 1979) pp. 95/ff. ServiceMaster is a committed evangelical Christian company, and Emery is now president of the Billy Graham Evangelistic Association.

12. Tetsuo Sakiya, *Honda Motor: The Men, The Management, The Machines* (Tokyo: Kodanshi International, 1982), p. 70. Honda made this comment in recalling an incident from the early days of his firm. A visitor from the United States accidentally dropped his false teeth into a primitive latrine. Honda personally climbed into the latrine and retrieved the teeth.

13. Lois Therrien, "The Rival Japan Respects," *Business Week* (November 13, 1989).

■ THE INFLUENCE ANSWERS

Question 1:

A = 0 points
B = 10 points
C = 6 points

Question 2:

A = 10 points
B = 6 points
C = 0 points

Question 3:

A = 0 points
B = 10 points
C = 6 points

Question 4:

A = 0 points
B = 8 points
C = 10 points

Question 5:

A = 0 points
B = 2 points
C = 10 points

Question 6:

A = 5 points
B = 10 points
C = 5 points

Question 7:

A = 10 points
B = 6 points
C = 3 points

Question 8:

A = 10 points
B = 0 points
C = 0 points

A Final Word

Not long ago I was in Atlanta taping a television interview on the creation of customer-driven companies. At the end of that session, the interviewer asked me, "Do you think this is a revolution?" I thought for a minute and said, "I'm not sure of the proper label, but I do know that it's tremendously important. It is significant enough that organizations are going to have to change in momentous ways. They're going to have to change their behavior, their attitudes, their structures, their compensation, their values. They'll have to change their very culture to make this work."

My company has worked with organizations who are struggling to become customer-driven. We have learned that it's not enough just to take your front-line people aside and tell them it helps to smile at customers. That will not solve the problem. If you're really going to be a player you must make a systematic, organized effort to change. The change is not incremental, it is transformational. It is moving from being one kind of organization to being a totally different kind of organization.

We've compiled a list of the transformational aspects to show what we're talking about. Becoming a customer-driven company means moving:

from motivation through fear and loyalty	to	motivation through shared vision.
from an attitude that says, "it's their problem"	to	ownership of every problem that affects the customer.
from "the way we've always done it"	to	continuous improvement.
from making decisions based on assumptions and judgment calls	to	doing it with data and fact-based decisions.
from everything begins and ends with management	to	everything begins and ends with customers.
from functional "stovepipes" where departments base decisions solely on their own criteria	to	cross-functional cooperation.
from being good at crisis management and recovery	to	doing it right the first time.
from depending on heroics	to	driving variability out of the process.
from a choice between participative *or* scientific management	to	participative *and* scientific management.

That's a lot of change to experience at one time. We're finding that organizations which can successfully navigate their way through this transformation have two key characteristics.

First, they start by saying, "This is significant, this is important, this is worth doing." Donald Peterson, former chairman of Ford, said, "It's a matter of life or death organizationally." Therein lies the threat and therein lies the opportunity.

Second, those who lead a change of such magnitude fully appreciate the enormity of the task, the challenges it will throw in the face of every unit, even every person, in their organization. With this profound appreciation in mind, they approach the change with a mixture of commitment and grace. The people in the organization who are the real leaders of the change—at all levels—are clear about their destination and persistent in their pursuit of it. But they also know that it will take time for others to embrace the changes as fully as they themselves have.

In general, people don't change easily or naturally. They need time; they need support. The sensitive leader knows they do and responds by giving people the space to struggle, even fail. He or she knows that if people fail it means they are trying, and that these failures are milestones on the road to success.

Let me summarize what we've said in two ways. First, I'd like to go back to the start of the book and revisit. Our purpose was to look at how companies can do right by the customer *every time*. I hope you now have a better idea of how to do that. More than that, I hope you've been able to identify ideas that, when implemented in your work environment, will produce a happy customer somewhere down the line. And I hope you'll come back to this book for more ideas as you continue your journey.

Second, I'd like to offer a quotation that I use in speeches. When I make a presentation, I always try to figure out a good way to close; it's the hardest part. I found a quotation that I like from Stanley Marcus of Neiman Marcus, the renowned retailer. I think he sums up what I have to say:

> There are only two things of importance. One is the customer, and the other is the product. If you take care of customers, they come back. If you take care of your product, it doesn't come back. It's just that simple. And it's just that difficult.

That says it well.

APPENDIX
TOOLKITS
ACKNOWLEDGMENTS
INDEX

Appendix:
The Research

The Forum Corporation has conducted widely varied research projects to identify the practices that result in a customer-driven organization.

In separate studies, we have examined:

- the relationship between managerial practices and employee attitudes on one hand and customers' assessment of quality on the other,
- the practices that result in customer-focused selling,
- how managers use influence to benefit customers,
- what leaders do to create customer-driven companies,
- the characteristics of organizations that succeed in focusing their resources on the customer.

One study, the Customer Focus Executive Assessment, conducted in 1989, is particularly relevant to this book. A survey questionnaire was administered in 44 companies to 563 senior executives. The research group developed a list of 84 organizational characteristics that had been widely believed to result in customer-focused organizations—practices defined by statements such as,

> We provide opportunities for employees at a variety of levels and functions to meet with customers.

Or,

> Executives demonstrate by their actions that customers' satisfaction is important.

The organizational characteristics on this initial list were chosen according to results of previous Forum Corporation research and a careful study of the literature on quality and customer service.

The senior executives answering the survey questionnaire were asked to do three things:

1. Rate their own organization for its achievement of customer-oriented performance outcomes, such as whether it typically exceeded customers' expectations, whether it provided value for the customer, and whether it achieved customers' loyalty.

2. Report essential financial data for their organization, such as return on investment and rate of annual increase in revenues.

3. Give their judgment about the extent to which their organization possessed each of the eighty-four characteristics.

Then the data were analyzed. The executives' estimates of whether their organizations achieved customer-oriented performance outcomes were compared with the financial data, and a high correlation was found: companies whose executives reported that they achieved customer-oriented performance did in fact achieve exceptional financial performance.

At the same time, the executives' estimates of the extent to which their organizations possessed each of the eighty-four characteristics was compared to their rankings of their organizations on customer-oriented performance outcomes. A ranking scale was developed: where an organizational characteristic was correlated with one of the customer-oriented performance outcomes at the level statisticians refer to as $p = .01$ or greater, that characteristic of the organization received credit on the ranking scale.

Further statistical analysis resulted in the forty characteristics that were, on the whole, most strongly correlated with good performance for the customer. These were grouped by the technique called factor analysis into seven categories.[1]

A list of forty characteristics in all seven categories appears on page 221 in a self-test that forms the first part of our Customer-Focus Toolkit. The lists of characteristics provide a useful check for any organization attempting to focus itself on the customer more tightly—an opportunity to test yourself and determine where your organization is strong and weak.

Among the forty-four organizational characteristics that did not demonstrate a strong correlation with favorable customer outcomes were:

- Employees are financially rewarded for being customer-focused.
- We use nontraditional work structures and methods, such as quality circles or self-managing work teams, in order to improve our results.
- We have specific goals for improving quality.

The failure of these characteristics to demonstrate a strong correlation with favorable outcomes for customers does not mean that they are necessarily unimportant. It is possible, for example, that some organizations in the sample were using nontraditional work structures with great success to achieve competitive advantage and others were using nontraditional work structures poorly. The overall result would be low correlation between use of nontraditional work structures and favorable outcomes for customers.

Where high correlations appear, however, that strongly suggests organizations need to strive to create those characteristics.

The correlations powerfully support all but one of the seven imperatives identified in this book. The exception is, "Measure, measure, measure." We believe that our analysis was unable to demonstrate a correlation between use of measurement and favorable outcomes for customers because most companies aren't measuring the right things in the right ways. In Chapter 6 we set out to show how to use measurement correctly.

▪ NOTE

1. Statistically, the seven categories are described as orthogonal to one another. The characteristics in each of the categories appear, according to statistical techniques, to correlate with each other: organizations that have one of the characteristics in each group are more than likely to have the others. But there is little or no correlation across the groups.

Customer-Focus Toolkit

On these pages are thirty-three tools that companies have found helpful in improving their customers' experience. Each tool has wide applicability, from factories to marketing departments to the office of the chief financial officer. The toolkit is divided into three sections:

1. The Characteristics of a Customer-Driven Company:
   A Self-Test
2. Tools for Developing a Vision
3. Tools for Smashing Barriers to Excellent Performance

Toolkit 1

■ THE CHARACTERISTICS OF A
CUSTOMER-DRIVEN COMPANY:
A SELF-TEST

Forum Corporation's Customer Focus Executive Assessment (see Appendix: The Research, 215–218) demonstrated that forty characteristics of corporations are highly correlated with success in meeting customers' needs. A technique called factor analysis showed that the characteristics break into seven clusters: The characteristics in each of the clusters are correlated with one another, meaning that organizations having one of the characteristics in each cluster are more than likely to have the others.*

All characteristics in each of the clusters are listed here. Use this list as a diagnostic self-test for your own organization. It will help you determine

*Statistically, the characteristics in each cluster are orthogonally related to one another.

where you most need to work and where your strengths lie. The discussion at the end of the test tells you which chapters in the book provide help with the problems you consider most serious.

For each characteristic, rate the extent to which the statement is true about your own organization, using this scale.

1—Not at all
2—To a small extent
3—To a moderate extent
4—To a great extent
5—To a very great extent

Then add up the scores for each cluster in the space entitled Your Score. Next, calculate your percentage rating by dividing your score by the highest possible score.

Vision, Commitment, and Climate

1. Our organization is totally committed to the idea of creating satisfied customers. ___

2. Rather than having to undo mistakes, we aim to "do things right the first time." ___

3. Executives demonstrate with their actions that customer satisfaction is important. ___

4. Our goal is to exceed the expectations of our customers in the things that matter most to them. ___

5. Being customer-focused is a major factor in determining who gets ahead in our organization. ___

6. Our organization is totally committed to the idea of quality. ___

7. Serving customers' needs takes precedence over serving our internal needs. ___

Your Score ___
÷ a possible 35 (your percentage score) = ___ %

Aligning Ourselves with Our Customers

1. When it comes to selling, we play a consultative or partnership role with our customers. ____

2. In our advertising and promotional materials, we avoid promising more than we deliver. ____

3. We know which attributes of our products or services our customers value most. ____

4. Information from customers is used in designing our products and services. ____

5. We strive to be a leader in our industry. ____

Your Score ____

÷ a possible 25 (your percentage score) = ____ %

Readiness to Find and Eliminate Customers' Problems

1. We monitor customer complaints. ____

2. We regularly ask customers to give us feedback about our performance. ____

3. Customers' complaints are regularly analyzed in order to identify quality problems. ____

4. We look for ways to eliminate internal procedures and systems that do not create value for our customers. ____

Your Score ____

÷ a possible 20 (your percentage score) = ____ %

Using and Communicating Customer Information

1. We know how our customers define "quality." ____

2. We provide opportunities for employees at various levels and functions to meet with customers. ____

3. We clearly understand what our customers expect of our organization. ____

4. We regularly give information to customers that helps shape realistic expectations.   _____
5. Our key managers clearly understand customers' requirements.   _____
6. Within the organization, there is agreement about who our "real" customer is.   _____
7. Our executives have frequent contact with customers.   _____

Your Score   _____

÷ a possible 35 (your percentage score) =   _____ %

Reaching out for Our Customers

1. We make it easy for our customers to do business with us.   _____
2. Employees are encouraged to go above and beyond to serve customers well.   _____
3. We try to resolve all customer complaints.   _____
4. We make it easy for customers to complain to us about our products and services.   _____

Your Score   _____

÷ a possible 20 (your percentage score) =   _____ %

Competence, Capability and Empowerment of People

1. We treat employees with respect.   _____
2. Employees at all levels have a good understanding of our products and services.   _____
3. Employees who work with customers are supported with resources that are sufficient for doing the job well.   _____
4. Even at lower levels of our organization, employees are empowered to use their judgment when quick action is needed to make things right for a customer.   _____

5. Employees feel they are involved in an exciting enterprise. _____

6. Employees at all levels are involved in making decisions about some aspects of their work. _____

7. Employees are cross-trained so that they can fill in for each other when necessary. _____

Your Score _____

÷ a possible 35 (your percentage score) = _____ %

Continuously Improving Our Processes and Products

1. Instead of competing with one another, functional groups cooperate to reach shared goals. _____

2. We study the best practices of other companies to get ideas about how we might do things better. _____

3. We work to continuously improve our products and services. _____

4. We systematically try to reduce our research-and-development cycle times. _____

5. When problems with quality are identified, we take quick action to solve them. _____

6. We invest in the development of innovative ideas. _____

Your Score _____

÷ a possible 30 (your percentage score) = _____ %

■ Analyzing Your Organization or Workgroup and Addressing Its Problems

After calculating your percentage score in each cluster, look at where you are high and where you are low. These suggestions will lead you toward opportunities for improvement:

If you're low in . . .	opportunities for improvement appear in . . .
Vision, commitment, and climate	Chapter 1 ("Create a Customer-Keeping Vision")
Aligning ourselves with our customers	Chapter 1 ("Create a Customer-Keeping Vision") and Chapter 2 ("Saturate Your Company with the Voice of the Customer")
Readiness to find and eliminate customers' problems *and* Using and communicating customer information	Chapter 2 ("Saturate Your Company with the Voice of the Customer")
Reaching out for our customers	Chapter 2 ("Saturate Your Company with the Voice of the Customer") and Chapter 5 ("Smash the Barriers to Customer-Winning Performance")
Competence, capability, and empowerment of people	Chapter 4 ("Liberate Your Customer Champions")
Continuously improving our processes and products	Chapter 3 ("Go to School on the Winners") and Chapter 5 ("Smash the Barriers to Customer-Winning Performance")

Chapter 6 ("Measure, Measure, Measure") is important for ensuring that you're truly achieving what you believe you're achieving.

Chapter 7 ("Walk the Talk") is vital to everyone who leads people in creating a high-quality experience for customers.

Toolkit 2

■ TOOLS FOR DEVELOPING A VISION

Every organization or workgroup needs a vision—a clear and exciting picture of what it seeks to become. And many individuals within workgroups can greatly stimulate their own performance and well-being if they develop their own personal vision.

In creating a vision, the goal is to summarize your ideal picture of the future in a concise, colorful statement. It can be very short: Founder Ray Kroc's vision for McDonald's was: "Quality, Service, Cleanliness, Value." He summarized the organization's goal in a formula that was impossible to misunderstand and hard to forget.

To create a vision for an organization or group:

1. Create a group to work on vision.
2. Use the tools in this section to help you think about vision.

3. Create a draft vision as a group.

4. Have all members of the group talk to others inside and outside of your organization to get feedback on the vision.

5. Redraft.

6. Repeat Steps 4 and 5 if necessary.

7. Promote the final vision throughout your company.

Here are six tools useful in thinking about a vision for an organization, workgroup, or individual. You may not need to use all of them, but try at least two or three. They'll help you start to envision what you *can* be.

1. **Imaginary journalism**—Imagine that you are a journalist writing an article for your favorite business publication. Create a story vividly describing the successes you and your workgroup will have achieved at a future time—two, five, or even ten years from now.

2. **Your values**—Think about what you and your workgroup value most. Then, list five ways of completing the phrase, "In my workgroup, we really care about. . . ."

3. **Customer wants**—Complete a paragraph that begins, "If I were a customer of my business unit, I would want. . . ."

4. **A picture**—Take a pencil or crayon and, on a blank sheet of paper draw a picture of how you want your workgroup to look in the future.

5. **Analogies**—Try shaping your vision with analogies. The leader of one sales group said: "I think of this group as a sports car—polished, tuned up, and ready to race." Listed here are categories that will prompt you to describe your workgroup with an analogy. For each category, write the image that comes to mind when you say, "If I were to describe my workgroup as a [*fill in category*], I would say it is. . . ." Categories to consider are:

Color—

Season—

Sport—

Geographic location—

Song or other music—

Movie—

Machine—

Emotion—

Food or beverage—

6. **The Five Whys**—Use a Five-Whys worksheet like this:

■ Five Whys Worksheet

Think of a work issue that is extraordinarily important to you. Then, use the "Five Whys" to discover key values associated with that issue.

Example: *The computer network is not reliable.*

1. Why is that important?

   The staff gets frustrated when the network isn't working.

2. Why is that important?

   I need to take time to help them solve the problem.

3. Why is that important?

   Projects don't get finished on time.

4. Why is that important?

   We can't deliver what we promised to our customers.

5. Why is that important?

   Customer satisfaction is our top priority.

Toolkit 3

TOOLS FOR SMASHING BARRIERS TO
CUSTOMER-WINNING PERFORMANCE

The tools in this section can help you and your organization manage with facts and data, and thus break down the barriers that prevent people from consistently doing right by the customer. These tools are particularly useful as part of the basic six-step procedure for breaking down barriers introduced in Chapter 5 (124–128).

Selecting the right tool for each problem depends not only on choosing tools that are *appropriate*, but also on your team's knowledge of, and previous experience in, applying each tool.

The chart beginning on page 231 lists each tool, some of its uses, and at which steps in the six-step procedure it is useful.

The chart also indicates the difficulty in using each tool:

● indicates that this tool is easy to learn. Generally, you can begin using tools marked with this symbol after simply reading about them in these pages.

■ indicates that a tool is more difficult to learn and apply correctly. You may be able to use these tools without specialized training from a professional, but before attempting to do so it's wise to study the books listed at the end of the discussion of the tool under the heading, "For further information."

◆ indicates that the tool is difficult to master. Enlist an expert or obtain specific training before attempting to use it.

Tool Selection Guide

Legend
● Easy to use
■ More difficult to use
◆ Most difficult to use

	Use this tool when . . .	Step 1 Collect info	Step 2 Convert to measures	Step 3 Analyze process	Step 4 Design improved process	Step 5 Establish standards	Step 6 Manage performance	Page
Benchmarking	You want to learn how your organization performs relative to other organizations (both competitors and noncompetitors) so you can continually match or exceed the work flow and work processes of your toughest competitors. This information is very useful in setting targets.	■	■		■			235
Brainstorming	You want to use an idea-generating technique that involves the spontaneous, unhindered contribution of ideas from all members of a group. This method is useful whenever you want to creatively explore options.		●	●	●	●		237
Cause-and-Effect Diagram	You want to analyze the current process, evaluating the interaction between two variables.			■				239
Charts and Graphs	You need to display collected data in a visual, effective, and easy-to-understand format.			■	■		■	243
Check Sheet	You want to record how often specified events occur so you can analyze the current process.	■					■	247

Tool Selection Guide

Legend
● Easy to use
■ More difficult to use
◆ Most difficult to use

	Use this tool when . . .	Step 1 Collect info	Step 2 Convert to measures	Step 3 Analyze process	Step 4 Design improved process	Step 5 Establish standards	Step 6 Manage performance	Page
Control Chart	You want to determine whether a process is statistically in or out of control.			◆	◆		◆	249
Cost-Benefit Analysis	You want to calculate the potential costs and benefits of a process improvement to determine whether it's truly worthwhile.				◆			251
Dimensions/ Expectations	You want to convert expectations, stated in customers' language, into specific dimensions that can be consistently measured.	●	●					253
Focus Group	You want to collect information from customers about which of your work unit's outputs have the highest priority for improvement, as well as suggestions for changes in existing products, services, or information or opinions on new products, services, or information.	◆						256
Force-Field Analysis	You want to understand the barriers that hinder change and the opposing actions that promote change. With this information, you can analyze the current process and design an improved process.			■	■		■	258

The Customer-Driven Company

Tool Selection Guide

Legend
● Easy to use
■ More difficult to use
◆ Most difficult to use

	Use this tool when …	Step 1 Collect info	Step 2 Convert to measures	Step 3 Analyze process	Step 4 Design improved process	Step 5 Establish standards	Step 6 Manage performance	Page
Histogram	You want to display data in ranked order to identify trends or tendencies.		◆	◆				260
Interview	You want to collect information from your customers, including information on current expectations and perceptions, ideas for improvement, or responses to your ideas.	●						264
Nominal Group Technique	You want to solicit many ideas from a group in order to convert customer information into measures to help establish a preliminary performance target. It also can be used to generate group information for mapping an improved process.	●	●	●	●			266
Pareto Chart	You need to define and prioritize the significant problems that may exist in the current process.			■	■			268
Process Mapping	You want to examine the flow of activities in a process along with related measures. This technique can also be used for designing a new process.			■	■			272
Process Evaluation	You must analyze the current process once it has been mapped.			■	■			274

Tool Selection Guide

Legend
● Easy to use
■ More difficult to use
◆ Most difficult to use

	Use this tool when . . .	Step 1 Collect info	Step 2 Convert to measures	Step 3 Analyze process	Step 4 Design improved process	Step 5 Establish standards performance	Step 6 Manage performance	Page
Run Chart	You want to find out where the breakdowns occur in the current process.		■	■			■	278
Sampling	You need to determine how many or how much of a population of items or people should be measured for the results to be valid. This tool also enables you to select the actual sample of the population you aim to study.	◆						281
Scatter Diagram	You want to examine what happens to one variable when another changes.	■			■			283
Standards Matrix	You need to establish standards for individual performers for each step in a process.					●		286
Stratification	You want to uncover hidden trends and potential causes of problems by separating data into meaningful categories.			◆				288
Survey	You are collecting information from large numbers of customers or other respondents.	◆	●					290
Taking Customer's Place	You want to gather first-hand information about your organization and the quality of your delivery of products, services, and information to customers.	●			●	●	●	293

The Customer-Driven Company

■ **Benchmarking**	**More difficult** ■

What   Special visits to or studies of other organizations are extremely helpful in breaking down barriers.

When   Use benchmarking to:
- identify the specific gaps between customers' expectations and your performance (Step 1).
- set preliminary improvement targets (Step 2).
- design an improved process (Step 4).

Who   Managers, perhaps with help from a marketing department or others with contacts outside the organization, and other employees skilled in the technique.

How   To conduct benchmarking:
- Refer to your customer data and determine expectations that are not being met. Use those expectations to establish the items to benchmark.
- Determine who has the "best-in-class" reputation, if you are benchmarking your direct competitors. If you are benchmarking noncompetitors, identify those who have an outstanding reputation in delivering the business processes you are benchmarking. Consider similarity of processes and products, nature of the work, and size of the organization.
- Collect information and data by direct contact with the organizations being benchmarked, whenever possible. This step includes face-to-face and telephone interviews, questionnaires, and surveys. In addition to direct contact, use professional contacts or consultants, technical journals, trade publications, advertisements, annual reports, or customer interviews. Offer to trade information if you make direct contact.

- Review information and data to determine the levels of success your competitors and non-competitors are reaching. Determine the top performers among your competitors and among noncompetitors for each benchmark item. For further information, see Chapter 3.

Results   Benchmarking provides:
- information about whether your direct competitors' performance is meeting or exceeding your customers' needs; if it is, your performance must be at least as good as that of your best direct competitor.
- information about noncompetitors' performance on each benchmarked item; if they exceed your customers' expectations and you do not, then you have an opportunity to improve your procedures to make them as good as those of your noncompetitors on each benchmarked item.

For further information   *Process Quality Management and Improvement* (Indianapolis, Ind.: American Telephone & Telegraph, 1989).

Beating the Competition: A Practical Guide to Benchmarking (Vienna, Va.: Kaiser Associates, 1988).

Leadership Through Quality: Implementing Competitive Benchmarking, Employee Involvement and Recognition (Stamford, Conn.: Xerox Corporation, 1987).

Robert C. Camp, *Benchmarking* (Milwaukee, Wis.: ASQC Quality Press, 1989).

What   A technique by means of which a group of people generates original ideas in an uninhibited atmosphere. In Phase 1 of the session, the focus is on quantity of ideas. In Phase 2, reviewing the list ensures that everyone understands all the ideas. In Phase 3, further review eliminates duplications, unimportant issues, and clearly unworkable ideas such as those violating legal requirements. Brainstorming sessions help every member of a group to contribute creatively to problem-solving tasks.

When   Use brainstorming to:

- convert customers' expectations and perceptions into units of measure (Step 2).
- analyze and evaluate the current processes (Step 3).
- map the inputs, outputs, flow of activities, and measures of the current process (Step 3).
- design an improved process (Step 4).
- establish standards (Step 5).

Who   Managers and other employees.

How   To use the brainstorming technique:

Phase 1

1. Write down on a flipchart or blackboard a statement of the problem or the subject being discussed.

2. Choose a person to facilitate and record ideas.

3. Write down every idea in as few words as possible—check with the contributor when paraphrasing. Do not interpret or change ideas.

4. Encourage creative, wild, and seemingly ridiculous notions. Members of the brainstorming group and the facilitator should never criticize or judge ideas and suggestions.

Phase 2

Review the list to ensure understanding by everyone in the group.

Phase 3

Review the list to eliminate duplication, unimportant issues, and clearly impossible proposals. Get consensus on any issues that may seem redundant or unimportant.

Make the sessions dynamic by keeping them short; five to fifteen minutes is best.

Results   Brainstorming provides:

- lists of ideas that can be critiqued and edited, prioritized, and rank ordered; for example, from most important to least important.
- creative solutions to problems based on everyone's input.

What   Also called a fishbone diagram or an Ishikawa diagram, it is a chart that resembles a fish skeleton, with the problem statement represented as the fish's head. The purpose of a cause-and-effect diagram is to identify probable causes of the problem summarized in the box at the fish's head.

When   Use a cause-and-effect diagram to:

- identify the possible causes of a problem by sorting and displaying them.
- analyze the current process (Step 3) by reviewing all potential factors that may cause a problem in a process in which the breakdowns occur.
- identify probable and root causes of a problem (Step 3).

Who   Managers in collaboration with employees.

How   To use a cause-and-effect diagram:

1. State the problem briefly in the box at the right of the diagram (see Figures 1 and 2).

2. Write the four main factors that may be causing the problem in the four boxes at the end of the fish's ribs. In most manufacturing organizations the factors are likely to include **manpower, machines, methods,** and **materials,** and in administrative or service companies, **policies, procedures, people,** and **equipment.**

3. Generate probable causes within each main factor by brainstorming (see page 237). Using the fewest possible words, write each cause on the horizontal "fishbones" that intersect the appropriate fish ribs. Be careful to write causes and not symptoms.

4. Continue exploring the chain of causes by asking, "Why . . . why . . . why . . ." until you have determined the root cause or at least a probable cause.

5. Circle the most likely causes—not symptoms—on the finished diagram.

Results   A cause-and-effect diagram provides:
- a determination of all causes of a problem.
- help in identifying the true cause of a problem, not just the symptoms.
- an illustration of the interplay between related causes.
- an opportunity to go to the next step and analyze information about causes, using tools such as interviews (page 264), the scatter diagram (see page 283), or process mapping (page 272).

For further   Kaoru Ishikawa, *Guide to Quality Control*
information   (White Plains, N.Y.: Unipub, 1988).

Process Quality Management and Improvement (Indianapolis, Ind.: AT&T, 1989).

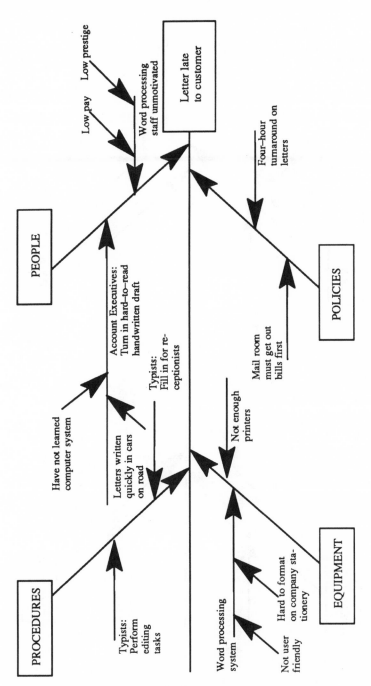

Figure 1   Cause-and-Effect Diagram: Late Letter

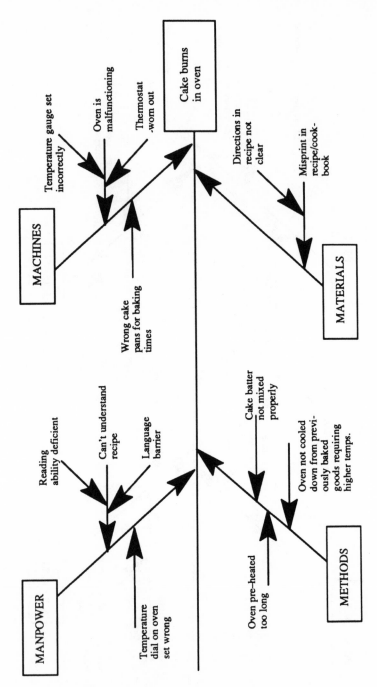

Figure 2   Cause-and-Effect Diagram: Burned Cake

The Customer-Driven Company

What   Several types of charts and graphs are useful for sum-
              marizing, organizing, and displaying data. These include
              line graphs, bar graphs, pie charts, and pictorial graphs.

When   Use charts and graphs to:
              - summarize the results of your data-collection
                efforts.
              - organize your data into an understandable and
                compelling form.
              - illustrate the results of data-collection exercises.
              - analyze the current process (Step 3).
              - design an improved process (Step 4).
              - manage performance (Step 6).

Who   Managers and anyone involved in collection and
              presentation of data.

How   To make most effective use of charts and graphs:
              - Label the chart according to what it depicts as a
                whole.
              - Label each axis according to what it represents,
                and include its unit of measure.
              - Choose an appropriate scale for each axis; the
                scale will be visually effective and can be deter-
                mined for maximizing differences (by choosing a
                small scale) or minimizing differences (by choosing
                a large scale).

              To determine which charts or graphs or both to use, first
              identify your objective.
              - **Line graphs** plot information and track perfor-
                mance. Label the axes with two variables for
                displaying pairs of related data, such as time and
                percentage of increase in production. Chart each

point that reflects one variable's relationship to the other. Data from several situations can be displayed on one chart or graph for comparison. This visual representation helps to show trends.

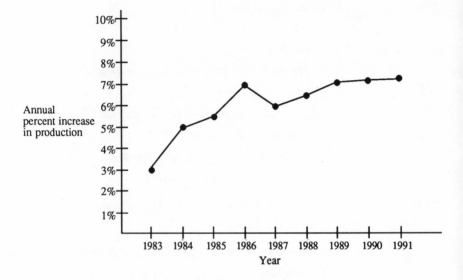

Figure 3   Line Graph: Yearly Production Increases

- **Bar graphs** demonstrate the quantities of given values and show relationships among those amounts. Bar graphs dramatize or minimize differences, depending on the scale of the vertical axis. Label the axes with two variables, then record the amount of one variable as it corresponds to the other.

The Customer-Driven Company

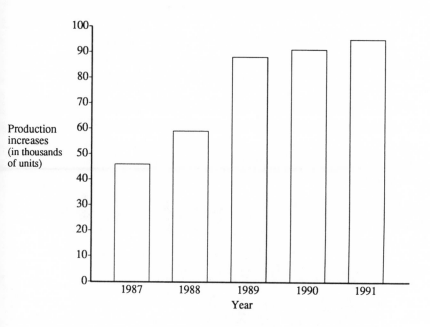

Figure 4   Bar Graph: Yearly Production Increases

- **Pie charts** are used to show value measured in relation to the whole. Considering the whole pie to be 100 percent, depict segments of the whole in measurable pieces that correspond to percentage values of the parts.

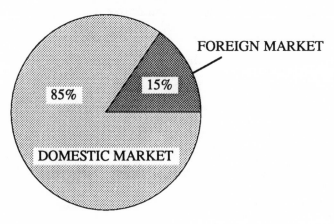

Figure 5   Pie Chart: Total Sales, 1990

- **Pictorial graphs** are used to depict, at fixed intervals, amounts of a variable in symbolic form. These graphs function much like bar graphs, except that pictures or symbols are used instead to fill the space to the appropriate amount or value. Be sure to indicate the nature and quantity of the variable represented by the chosen symbol.

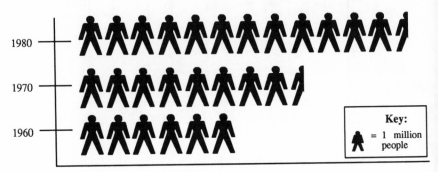

Figure 6 Pictorial Graph: Population Growth in Urban Areas

Results Charts and graphs provide:
- a mechanism for presenting data in a form that can be widely communicated.
- more information and greater effectiveness than words alone can deliver.

Note: Your charts or graphs should appear near the text that refers to them, or be shown in presentations in conjunction with corresponding text.

For further information Kaoru Ishikawa, *Guide to Quality Control* (White Plains, N.Y.: Unipub, 1988).
William Scherkenbach, *The Deming Route to Quality and Productivity* (Washington, D.C.: CEEP, 1988).

The Customer-Driven Company

What   A form constructed and used for recording how often an
                    event occurs.

When   Use check sheets when you need to:
- physically track the number of occurrences over a specified time of an event.
- understand the significance of the number of times an event occurs.
- collect information (Step 1).

Who   Both managers and other employees.

How   To use a check-sheet form:

1. Have everyone who will be collecting the data agree about what is being observed and counted. To ensure accuracy, practice collecting data so that all data collectors or observers understand the categories. **Caution:** Avoid using a category labeled "other problems."

2. Determine the period for which observations will be made; such as the month of May, the next peak workload period, the first quarter of fiscal 1992.

3. Develop a checklist form that is easy to understand and use. Columns and headings should be clearly marked (see Figure 7) and leave enough space for tally marks or check marks.

4. Collect data consistently.

5. Count the occurrence of the event being observed over a specified period.

Results   A check sheet provides data that can be used to:
- track defects in items or processes.
- detect patterns that signal underlying problems.

Customer Complaints	October				
	Week 1	Week 2	Week 3	Week 4	Total
Error on statement	ЖГ ЖГ ЖГ ЖГ I	ЖГ ЖГ ЖГ ЖГ IIII	ЖГ ЖГ ЖГ ЖГ ЖГ I	ЖГ ЖГ ЖГ ЖГ I	92
Late statement	ЖГ ЖГ	ЖГ I	ЖГ	ЖГ II	28
ATM not operating	ЖГ ЖГ ЖГ ЖГ ЖГ ЖГ ЖГ	ЖГ ЖГ ЖГ	ЖГ ЖГ ЖГ ЖГ ЖГ ЖГ ЖГ II	ЖГ ЖГ ЖГ ЖГ ЖГ ЖГ	117
Total	66	45	68	58	248

Figure 7   Check Sheet: Customer Complaints

- identify a logical starting point in the problem-solving cycle.
- construct Pareto charts (see page 268).
- construct histograms (see page 260).

For further information   Kaoru Ishikawa, *Guide to Quality Control* (White Plains, N.Y.: Unipub, 1988).

What   A statistical method for understanding how a process
                 functions and for monitoring variation in that process.
                 Specifically, control charts are used to determine whether
                 a process is, statistically speaking, in or out of control.
                 They include upper and lower "control limits" that can
                 be calculated with prescribed formulas; then, individual
                 units of measure can be placed on the chart to determine
                 if they fall above or below a specified range or average.
                 Data points falling below the lower control limit or rising
                 above the upper limit can provide information about the
                 incidence of variation. **Note:** Control charts detect only
                 changes or variation in a process; alone, they do not
                 determine the cause of the variation. A control chart can
                 help you determine if the process is out of control, so that
                 you may then take necessary action.

When   Use control charts to:
                 • analyze and evaluate the variation in the current
                   process (Step 3).
                 • design an improved process (Step 4).

Who   The person in charge of the process being monitored,
                 usually with help from someone expert in statistics.

How   To use control charts:
                 • Check if all points in the sample are within the
                   control limits and whether the points are taking a
                   recognizable form or pattern.

                 If all points are within the control limits and no patterns
                 appear; then the process is in control, and any
                 improvements will further reduce the variation.

If a process is found to be out of control, the causes should be investigated and the variation in the process reduced to bring it within control limits.

If patterns appear within the control limits, such as clusters of points above or below the central line, a discernible rise-and-fall configuration, or other pattern, some abnormality may be occurring and should be investigated.

Results

A control chart can be used to:

- make decisions based on data.
- help affect a process and improve it.
- provide a determination of whether a process is in or out of control, signaling that the user should take such actions as investigating, eliminating, or continuing to monitor the process.

For further information

H. James Harrington, *The Improvement Process: How America's Leading Companies Improve Quality* (New York: McGraw-Hill, 1987).

Kaoru Ishikawa, *Guide to Quality Control* (White Plains, N.Y.: Unipub, 1988).

William Scherkenbach, *The Deming Route to Quality and Productivity* (Washington, D.C.: CEEP, 1988).

■ Cost-Benefit Analysis   Most difficult ◆

What An analysis that compares the expense of a proposed improvement to the potential savings and increased benefits (revenues and so on). The analysis answers the questions, "What will it cost?" and "What is the return on the investment?"

When Use cost-benefit analysis to:
- choose among several possible process improvements (Step 4).
- determine the worth of a proposed improvement.
- identify all the benefits of a proposed solution (even those which are not obvious).

Who Managers and other employees.

How To use cost-benefit analysis:
- Calculate the known costs of the proposed improvement. Devote time to thinking of additional costs you may have forgotten.
- Calculate the potential benefits of the proposed improvement. Benefits include such changes as increased productivity, decreased rework, higher profits, reduced costs due to reduced man-hours, and intangibles such as improved customer satisfaction and quality of work life.
- Subtract the costs from the benefits. The remainder will be the objective of the analysis.

It is sometimes necessary to project several years from the date of implementing the improvement to amortize the investment made in the cost category. For example, an improvement team might recommend a new piece of test equipment to help reduce rework. The cost of the

3

equipment, $20,000, includes wiring, installation, and training. The team calculates the benefits as $10,000 in the first year and $6,000 for each of the three years thereafter. Thus, a simplified cost-benefit analysis might calculate the net benefit after four years to be $8,000.

A more sophisticated cost-benefit analysis would discount the value of benefits to be achieved in future years by an appropriate percentage. Using a 10% discount rate, for example, the $10,000 benefit in the current year is still worth $10,000, but the value of $6,000 in benefits in year 2 is $5,400; the value of $6,000 in year 3 is $4,860; the value of $6,000 in year 4 is $4,374. The net present value of all the benefits totals $24,634, and thus the net benefit of the improvement is $4,634.

Results A cost-benefit analysis can be used to:

- justify expenses related to proposed improvements and continuous and profitable improvement of the process.
- help set criteria for implementing suggestions.

For further information David A. Garvin, *Managing Quality* (New York: Free Press, 1988).

■ Dimensions of Customers'
Expectations and Perceptions   Easy ●

What   A list that categorizes customers' expectations and
                  perceptions and provides a framework for categorizing
                  information from customers.

When   Use the dimensions of customers' expectations and
                  perceptions to:
                  - identify your customers' expectations about
                    outputs of your work unit.
                  - identify the specific gaps between customers'
                    expectations and perceptions in order to prioritize
                    opportunities for improvement (Step 1).
                  - convert measures (Step 2).
                  - set up the exercise of taking the customer's place
                    (see page 293).

Who   Managers and employees.

How   To use the dimensions of customers' expectations and
                  perceptions, review the list below* and select categories
                  that apply to your customers' expectations and percep-
                  tions. Add any dimensions that you feel pertain specifi-
                  cally to your customers' situations. (**Note:** Throughout
                  this list, *output* is used to represent your products,
                  services, or information.)
                  - **Performance:** how well the output does what it is
                    supposed to do.
                  - **Reliability:** the ability of the output (and its
                    provider) to function as promised.

*The first eleven dimensions of customer expectations listed in this section are
based on work by D. A. Garvin (Harvard Business School) on factors of product
quality and A. Parasuraman, V. A. Zeithaml, and L. L. Berry (Texas A&M) on
factors of service quality.

- **Assurance:** the knowledge and courtesy of the work unit's employees and their ability to elicit trust and confidence.
- **Tangibles:** the physical aspects of your output.
- **Empathy:** the demonstration of caring and individual attention to customers.
- **Responsiveness:** willingness and readiness of employees to help customers and provide prompt service.
- **Features:** the characteristics of your output that exceed the output's basic functions.
- **Conformance:** the degree to which an output meets specifications for design and operation.
- **Durability:** how long your output lasts.
- **Perceived quality:** the relative worth of your output in the eyes of customers.
- **Serviceability:** how easy it is for you or the customer to fix your output with minimal downtime or cost.
- **Cost:** the value offered by your output relative to its price.
- **Choice:** the options you provide.

Results   Dimensions of customers' expectations and perceptions provide:

- a list of categories for organizing and measuring information from and about your customers, so that you can obtain a clear picture of what your customers care about, where you stand in their eyes, and what to improve upon to meet their expectations.

The Customer-Driven Company

- a framework to keep in mind when carrying out the exercise of taking the customer's place to learn his or her perspective.

For further information

David A. Garvin, *Managing Quality* (New York: Free Press, 1988).

A. Parasuraman, V. A. Zeithaml, and L. L. Berry (Texas A&M). *SERVQUAL: A Multiple-Item Scale for Measuring Consumer Perceptions of Service Quality* (Cambridge, Mass.: Marketing Sciences Institute, 1986).

3

What   A group of twelve or fewer external or internal customers led by a moderator in a group-interview format. Focus groups provide a forum for discussions, ideas, and feedback. Discussions may be recorded (with permission) by a secretary or with audio- or videotape.

When   Use focus groups to:
- collect information from customers about their expectations, perceptions of what they receive, and suggestions for improvement (Step 1).
- find out from customers which outputs of your work unit have the greatest need for improvement (Step 1).
- understand how customers think.
- obtain constructive criticism.

Who   Managers trained as facilitators or other skilled facilitators should lead the focus group.

How   To conduct a focus group, with the help of a moderator or a facilitator:

1. Prepare the questions or discussion guide and distribute it ahead of time to people participating in the focus group (bring extra copies on the day of the meeting).

2. Determine how many and which customers to invite, based on the total customer group, to ensure accurate, fair representation (see guidelines for sampling, page 281).

3. Design a plan for analyzing the results.

4. Assemble the group, conduct the discussion, organize input, and report results.

Results	A focus group provides:
	• the possibility of uncovering a wide range of customers' perceptions.
	• a rapidly generated list of ideas for improving service and solving problems.
	• in-depth and credible information from customers.
	• information that can be used to demonstrate the need for improvement.
For further information	David L. Morgan, *Focus Groups as Qualitative Research* (Newbury Park, Calif.: Sage Publications, 1988).

3

■ Force-Field Analysis   More difficult ■

What A technique for understanding a struggle between opposing forces. *Driving forces* move a situation toward change and *restraining forces* hinder change; if the latter force is the more powerful, change will not occur; it can take place only when these opposing forces have been modified. Force-field analysis helps to identify the forces that are opposing positive change.

When Use force-field analysis to:
- analyze how the current process works (Step 3).
- design an improved process (Step 4).
- establish standards that may require overcoming some restraining forces.
- develop rewards and recognition programs that are new and may face resistance (Step 6).
- encourage change by defining the organizational factors that either drive or block desired changes.

Who Managers or facilitators with groups of employees.

How To conduct a force-field analysis:
- Start by defining the current problem (perhaps with a cause-and-effect diagram) and the desired change or goal (see Figure 8).
- Identify the processes that either drive or restrain achievement of that goal and record them in the form of a balance sheet, with driving forces on the left and restraining forces on the right. (**Note:** Brainstorming and group-consensus techniques can be used to identify the opposing forces and their relative strengths or weaknesses.)
- Have group members evaluate and arrive at a consensus on analysis of the problem(s). For

example, agreement should be reached on the major restraining forces so that they can be worked on.

Results   A force-field analysis provides:

- a visual framework for examining opposing forces.
- a basis for planning and effectively implementing change; it is usually more effective and less threatening to eliminate or diminish the restraining forces than it is to increase driving forces.

For further information   H. James Harrington, *The Improvement Process: How America's Leading Companies Improve Quality* (New York: McGraw-Hill, 1987).

Mary Walton, *The Deming Management Method* (New York: Putnam Press, 1986).

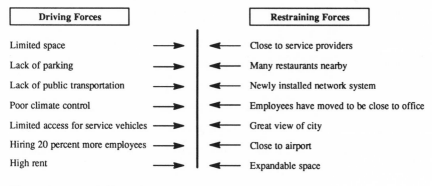

Driving Forces		Restraining Forces
Limited space	→ ←	Close to service providers
Lack of parking	→ ←	Many restaurants nearby
Lack of public transportation	→ ←	Newly installed network system
Poor climate control	→ ←	Employees have moved to be close to office
Limited access for service vehicles	→ ←	Great view of city
Hiring 20 percent more employees	→ ←	Close to airport
High rent	→ ←	Expandable space

Figure 8   Force-field Analysis: Should the Office Be Moved?

3

What

A bar graph that displays numerical information about the frequency of distribution of continuous data.

When

Use histograms when you need to:
- convert customer information into measures (Step 2).
- analyze process (Step 3).
- display data in ranked order.
- understand the distribution of data.

Who

Managers and other employees.

How

To create a histogram:
- Collect data (numerical) on the number of occurrences, errors, or other phenomena observed with respect to the problem you are trying to assess, and count the data points. For instance, in Table 1 and in Figure 9 on page 263, the number of employees taking one or more sick days was 38.
- Next, determine the range (R) for your group of data. In our example, the range is the difference between the smallest data point (fewest days) and largest data point (most days). Decide how many columns (bars) your histogram will have (usually between 6 and 12) and divide the range by the number of bars to determine how many intervals each bar in the histogram will cover. In Figure 9, the first bar, with an interval of 1 to 3 sick days, includes three employees (cases).
- Decide on the scale for the vertical axis. In Figure 9, the interval with most cases was chosen to define the upper limit of the range, with intervals of 2 creating a visually effective graphic display. You

may, however, use scales and intervals of 5 or 10 units or whatever seems logical and appropriate.

- Draw each bar.
- Calculate the measures of central tendency; in situations with extensive data, the natural tendency is for most cases to fall in the middle, at or near the average. The three measures of central tendency are:

    — The **average**, which is the sum of measured or counted data divided by the number of data points. In Table 1, employee sick days, 538, divided by the number of employees, 38, gives an average of 14.16 sick days per employee taking at least one sick day.

    — The **mode**, or the most commonly observed value, class, interval, grouping, or data point. In Figure 9, the mode is the interval of 19 to 21 sick days, into which range eight employees fitted.

    — The **median**, or the midpoint in a distribution of data. This is the point above and below which 50 percent of the data fall. In Table 1, the median is 15, because 50 percent of the data lie above and 50 percent below this midpoint.

- Add markings of the measures of central tendency to the histogram (see Figure 9). Use the measure(s) that make your point most dramatically. Be especially suspicious of your data if the tallest bar in your histogram occurs at the far left (positively skewed, as in Figure 10) or far right (negatively skewed, as in Figures 9 and 11).

Results   A histogram helps to:

- indicate the need for a solution that will get the process under control.
- determine in which order to work on problems.

For further information   Kaoru Ishikawa, *Guide to Quality Control* (White Plains, N.Y.: Unipub, 1988).

Mary Walton, *The Deming Management Method* (New York, N.Y.: Putnam Press, 1986).

Table 1   Number of Sick Days Taken (Data Points)

Number of sick days taken per employee per year	Number of employees	Number of sick days taken per employee per year	Number of employees	Number of sick days taken per employee per year	Number of employees
1	1	11	1	21	3
2	0	12	2	22	2
3	2	13	0	23	1
4	2	14	2	24	1
5	1	15	3	25	0
6	1	16	2	26	1
7	1	17	3	27	0
8	2	18	0	28	0
9	1	19	1	29	0
10	1	20	4	30	0

The Customer-Driven Company

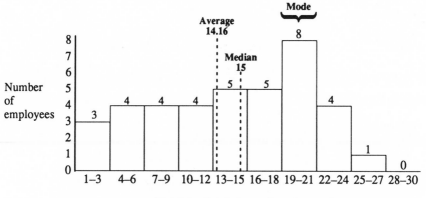

Number of sick days taken per employee

Figure 9   Histogram: Number of Sick Days Taken per Employee

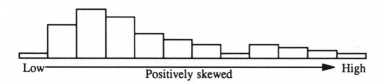

Low — Positively skewed → High

Figure 10   Histogram: Positive Skewness

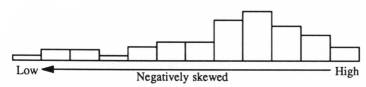

Low ← Negatively skewed — High

Figure 11   Histogram: Negative Skewness

What   A face-to-face communication with the customer or a
             significant participant in a process. A good interview
             includes well-chosen questions and active listening to the
             responses.

When   Conduct interviews to:

- collect information from customers (Step 1),
  particularly qualitative and anecdotal data.
- learn from customers which outputs of your work
  unit need improvement most (Step 1).
- identify specific customer expectations and
  perceptions (Step 1).
- gain personal understanding of customers' needs
  and generate ideas for improvement.
- test the waters when introducing new products,
  services, or changes that will affect customers.
- test the wording or focus of a survey being
  developed.

Who   Managers and other employees.

How   To prepare for an interview or set of interviews:

- Review background information on both the topic
  and the person to be interviewed.
- Determine areas to be covered in the interview.
- Write the questions; open-ended
  questions—asking for more than a yes or no
  answer—produce more information than closed-
  end questions.
- Practice asking the questions with a colleague and
  refine as necessary.
- For consistent collection of data, adhere to the
  prepared protocol or script in all interviews.

- For best results, choose customers to be interviewed according to the sampling guidelines on page 281.

To conduct the interview:

- Establish the purpose of the interview and the benefit to the interviewee of giving his or her time. You might say, "We are interviewing customers to gather more information about how to improve our product so that it best meets your requirements." Then ask the customer for some of his or her time.
- Tell the customer how long the interview will take.
- Ask the interview questions and listen carefully to the customer's perspective.
- Follow up on vague responses by asking for further clarification.
- Take notes or record the customer's responses.
- Thank the customer for his or her time and contribution.

Results

An interview with a customer provides:

- understanding of the range of perceptions on an issue.
- a list of ideas for enhancing value for customers or for solving problems.
- new insights into people's expectations and problems.
- assistance in planning for improved delivery to customers.

What   A technique for providing all group members an equal voice in generating ideas. This is a highly structured form of brainstorming.

When   Use the nominal group technique to:
- collect information (Step 1).
- convert customers' information into measures (Step 2).
- evaluate the process by identifying where break-downs occur (Step 3).
- design an improved process (Step 4).

Who   Managers, other employees, and a skilled facilitator.

How   To conduct the nominal group technique:

1. Clearly identify, in writing, the topic for the group. The topic is best expressed as a question.

2. Have all team members respond to the question by writing down ideas they feel are most important.

3. Go around the group sequentially and solicit one idea at a time. Write these on a blackboard or a flipchart where everyone can see them. Get everyone's input. A member of the group who does not have an idea can "pass." To ensure that ideas are not judged or evaluated, this procedure must be closely monitored by the facilitator.

4. Make sure each idea appears on the blackboard only once. If a team member restates previous input in slightly different words, merge the two versions.

5. Have each team member copy the completed list of ideas on a sheet of paper, lettering and grouping them from A to Z, for convenience.

6. Have participants rank the ideas by importance, assigning the highest number to the most important idea;

if there are six ideas, ranking will be from 1 to 6, with 6 being the most important and 1 the least. For example, if a participant thought Safety was the most important of six ideas, his or her ranking might look like this:

A. Job security: 5
B. Safety of workplace: 6
C. Cleanliness of workplace: 3
D. Quality of products: 2
E. Money for salaries: 1
F. Advancement opportunities: 4

7. Collect participants' rankings.

8. Add the collective rankings for each idea; the idea with the highest total score is the one considered most important to the overall team. In the preceding example, safety of the workplace might earn the highest overall score from the group of six participants:

A. Job security   5,5,6,4,1,2 = 23
B. Safety of workplace   6,6,4,5,5,5 = 31
C. Cleanliness of workplace 3,1,3,6,6,4 = 23
D. Quality of products   2,2,1,1,4,6 = 16
E. Money for salaries   1,4,5,3,3,3 = 19
F. Advancement
   opportunities   4,3,2,2,2,1 = 14

Results   The nominal group technique provides:

- a rank-ordered selection of ideas.
- a procedure by which each member of a group has input into determining the whole group's priorities.

For further information   *Process Quality Management and Improvement* (Indianapolis: AT&T, 1989).

James H. Harrington, *The Improvement Process: How America's Leading Companies Improve Quality* (New York: McGraw-Hill, 1987).

What	A form of vertical bar graph that helps to identify problems by the frequency of their occurrence in a process. As a graphic display, the Pareto chart draws attention to, and enlists cooperation in, making improvements. The Pareto chart is effective because of its ability to graphically demonstrate how seemingly small matters can cause big problems; taller bars represent more significant problems and the shorter bars, less significant problems.

When Use Pareto charts to:

- categorize data to identify opportunities for improvement (Step 3).
- rank improvement opportunities and set objectives (Step 3).
- show the relative importance of problems (Step 4).
- assess conformity to customer requirements.
- improve process quality.

Who Managers, often in collaboration with and receiving input from employees.

How To construct a Pareto chart:

- Select problems to be compared and rank ordered by nominal group technique (see page 266) or by examining existing data (such as previous quality reports).
- Select the unit of measurement that appropriately quantifies problems to be assessed.
- Determine how much time to allow for collecting data.
- Draw horizontal and vertical axes on graph paper (see Figure 12). On the left-hand vertical axis, label the measurement values in equal increments.

- Draw in the bars, with the height of each bar determined by the corresponding value on the vertical axis. Each bar should have the same width and be drawn in contact with the bars next to it.
- Label each bar below the horizontal axis according to the problem it represents.
- Reorder the bars, going from left to right in order of decreasing frequency or cost.
- Label the vertical axis on the right-hand side of the graph as the cumulative percentage of the total distribution.
- Plot a percentage line showing the cumulative total reached with the addition of each problem category. Once all problems have been represented, the total distribution should be 100 percent.
- Title the graph and write the source of the data on which the graph is based; with quality control, the source of the data must be clear. Also, include all pertinent facts that will help define the parameters of observation.
- Compare the frequency or cost of each problem category relative to all others.

Results

A Pareto chart provides:

- an analysis of how a small percentage (typically 20 percent) of the problems cause a large percentage (typically 80 percent) of the defects or costs.
- a graphic demonstration of how much damage can be caused by a few vital errors and how much progress would be achieved if the "vital few" key errors were eliminated.
- a first step in making improvements.

Note: It is most worthwhile to work first on the problem (or cause) represented by the tallest bar on the Pareto chart. But, remember that the most frequent problems are not always the most costly.

For further information

Kaoru Ishikawa, *Guide to Quality Control* (White Plains, N.Y.: Unipub, 1988).

William Scherkenbach, *The Deming Route to Quality and Productivity* (Washington, D.C.: CEEP, 1988).

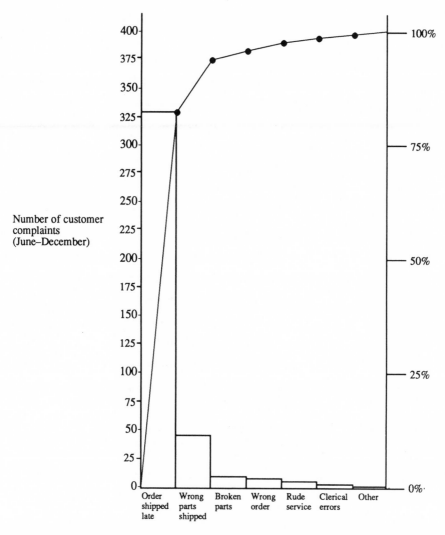

Figure 12   Pareto Chart: Cost to Correct Customers' Complaints

Process Mapping

What   A technique for examining a process to determine where and why major breakdowns occur. Process mapping is a first step in evaluating a process and then designing an improved one. In addition to illustrating how a process currently works, process mapping uncovers specific problems in the process so that improvements can be made. Once problems are understood, process mapping can also be used to map a new, improved process.

When   Use process mapping to:
- map the inputs, outputs, flow of activities, and measures of the current process (Step 3).
- evaluate the current process:
  - How well is the process working?
  - What are the breakdowns?
  - What are the probable causes?
  - What is a preliminary improvement target (Step 3)?
- design an improved process (Step 4).

Who   Managers across and within functions, in collaboration with groups of employees who are involved with the process.

How   To create a process map:

1. Construct a process-map worksheet such as Figure 13.

2. Identify the "key process": Using information from your customer, identify the process most responsible for a major gap between your customers' expectations and their perceptions of your performance.

3. Identify the major work units involved in the process and list them on the left-hand side of the process

map. Six work units are usually the maximum needed to provide sufficient detail.

4. Identify the starting point and enter that at the far left side of the map in the row of the appropriate function. As you move toward the right, enter the activities that are associated with each function. Avoid details. **Note:** The passage of time is indicated by the flow of the process from left to right.

5. Connect activities with an arrow from the supplier to the immediate internal customer. Arrows may go in both directions, indicating interaction between two functions related to the same activity, such as rework. Arrows may also go from one supplier to two or more internal customers, or from two or more suppliers to one customer.

6. Identify measures that exist for each output, once the entire process has been mapped. State these measures in a few words and place them next to the arrow that flows from the supplier to the customer.

The final diagram should be a record of how the current process actually operates. You may need several attempts or adjustments before the map is correct. See Figure 14.

Caution: Start by mapping the process as it really is, not as you think it should be. An accurate map of what's *currently* happening is an essential tool in creating improvements.

Results   Process mapping provides:

- focus on the connections and relationships between work units.
- a picture of all of the handoffs, activities, tasks, steps, and measures in a process.

Key Process: _____   Improvement Target: _____

→ → → → Time → → → →

Functions	Sequence of Activities and Measures

Date: _____

Figure 13   Process Map

- understanding of how various work activities are connected and where connections may be breaking down.

Process Evaluation

What   A technique for uncovering the probable causes of problems in a process and determining if the process should be changed or eliminated. A problem in a process is detected when a gap appears between desired and actual performance.

When   Use process evaluation to:
- analyze the current process (Step 3).
- evaluate the process:
  - How well is the process working?
  - What are the breakdowns?
  - What is the preliminary improvement target (Step 3)?

Example

Key Process: PRODUCE A DOCUMENT (CURRENT PROCESS)   **Improvement Target:** _____

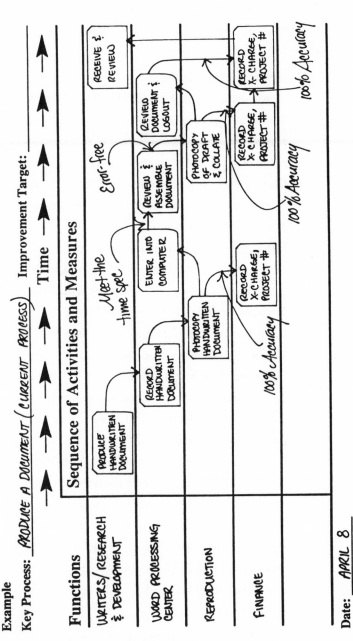

Figure 14   Process Map Completed

Date: APRIL 8

Who	Managers across and within functions, in collaboration with groups of employees who are involved with the process.
How	To perform a process evaluation:

- Construct a process map (see Figures 13 and 14).
- Construct a process-evaluation chart similar to Figure 15 using the information from the completed process map as a basis.
- Choose the most problematic process steps and record them in the boxes directly under the label Process Steps.
- Enter, in the boxes to the right of Actual Performance, a brief description of the actual, current performance for each step you identified.
- Enter, in the boxes to the right of Desired Performance, a brief description of how you would like to see that step performed.
- Describe, in the boxes to the right of Difference, the differences between actual and desired performance, if any; these differences may be measurable or perceived.
- Record, in the boxes to the right of Probable Cause, what you believe may contribute to the problem. Be sure to focus on the underlying problem, not the symptom; ask why the problem exists, not once, but five times, as in the Five Whys Worksheet on page 228, or construct a cause-and-effect diagram. These tasks will help to uncover the probable cause.

Results	A process evaluation helps to uncover the probable causes of problems in a process so that improvements can be made.

The Customer-Driven Company

Example

Process: _PRODUCE A DOCUMENT_   **Date:** _APRIL 28_

For the most problematic steps of the process, complete the following evaluation:

Process Steps

Factors For Evaluation	PHOTOCOPY HAND-WRITTEN DOCUMENT	REVIEW + ASSEMBLE DOCUMENT	REVIEW DOCUMENT & LOGOUT/DONE			
Actual Performance	4 HOURS	IN THE QUEUE A LONG TIME — 6 HOURS	THE DOCUMENT USUALLY DELAYED FOR 2 HOURS			
Desired Performance	1 HOUR	2 HOURS	NO LOST TIME — 30 MINUTES MAXIMUM			
Difference	3 HOURS	4 HOURS	TIME — 1½ HOURS			
Probable Cause	REQUIREMENT TO COPY EVERY NEW DOCUMENT	SUPERVISOR REVIEWS EVERY DOCUMENT	NO PROCEDURES CURRENTLY IN PLACE			

Figure 15   Process Evaluation

■ **Run Chart**   **More difficult** ■

What   A chart, sometimes called a trend chart, used to identify
              meaningful shifts in the average of data collected over a
              specified time.

When   Use run charts to:
              • understand trends in data.
              • evaluate trends in process performance:
                — What is the trend?
                — What are the breakdowns?
                — What is a preliminary improvement target?
              • visually represent data to determine trends in
                performance.
              • convert to measures (Step 6).
              • analyze process (Step 3).
              • manage performance (Step 6).

              Caution: A run chart shows every variation as potentially
              important. After making the chart you must identify
              which trends or shifts in the average are meaningful.

Who   Managers and other employees.

How   To construct a run chart:
              • Draw the vertical axis of the graph, which
                represents measurement of some variable.
              • Draw the horizontal axis of the graph, which
                typically represents time or sequence.
              • Plot the data on the graph.
              • Connect data points for easy use and interpreta-
                tion of the chart.
              • Clearly mark the time range, measurement range,
                and units of each.

The Customer-Driven Company

Collected data must be kept in the order in which they were gathered. Because a characteristic is tracked over time, the sequence of data points is critical.

Results A run chart can provide:

- results that can be used to monitor a process, which should have nearly equal numbers of data points above and below the average; if these numbers are far from equal, investigate to determine the cause of the variation.
- information about favorable shifts in activity, enabling you to make a permanent change in the process.
- information about unfavorable shifts in activity, prompting you to make necessary corrections in the process.

For further information Kaoru Ishikawa, *Guide to Quality Control* (White Plains, N.Y.: Unipub, 1988).

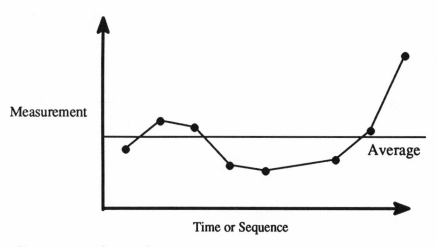

Figure 16   Sample Run Chart

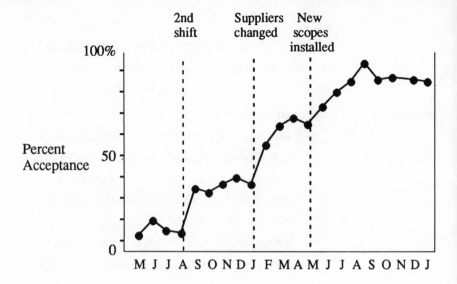

Figure 17   Run Chart: Percentage of Acceptance of Printed Circuit Boards on First Test

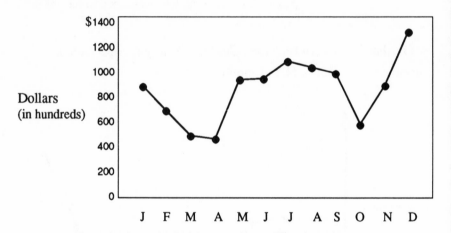

Figure 18   Run Chart: Family Expenditures per Month

The Customer-Driven Company

What   A technique for determining how many or how much of
                  an overall population should be tested or measured in
                  order to gain valid information from a study. Sampling
                  also refers to the actual selection of the representative
                  group.

When   Sampling is used to obtain accurate information about a
                  large group whose members are similar enough that every
                  item need not be observed or tested to understand the
                  varying characteristics of the entire population.
                  - collect data from customers (Step 1).
                  - determine how many people to interview, survey,
                    or include in a focus group (Step 1).
                  - determine the number of observations to include
                    in a run chart or a scatter diagram.

Who   Anyone who is collecting data and needs to know how
                  many observations will be enough and how to choose a
                  representative group.

How   To conduct sampling:
                  1. Identify the population you aim to study.
                  2. Choose a sample size large enough to accurately
                  represent the entire population. A person knowledgeable
                  in statistics can be helpful with this choice.
                  3. Determine which type of sampling is most
                  appropriate, based on how uniform the population is.
                  These types are:
                  — **Random sampling:** Decide which members of the
                    population will be included in the sample by
                    using a table of random numbers (available in
                    any statistics textbook).

3

— **Stratified sampling:** Examine the population for variations along some criteria, and choose a sample that reflects the population's distribution along those variables. For example, if a population of cars is 50 percent comprised of blue cars, 30 percent of red, and 20 percent of yellow, a sample of 10 cars should reflect this color distribution and therefore include 5 blue, 3 red, and 2 yellow cars.

— **Stratified random sampling:** Combine the two techniques. Stratify first, then take a random sample from each of the subpopulations. This sequence may work best where data come from a large number of sources.

Results Sampling is an efficient way of collecting data about a population of people or items without evaluating every member of that population.

For further information Kaoru Ishikawa, *Guide to Quality Control* (White Plains, N.Y.: Unipub, 1988).

What   A graphic display that illustrates what happens to one variable or process when another variable or process changes. Scatter diagrams are also used to test for possible cause-and-effect relationships.

When   Use scatter diagrams to:
- study the possible relationship between one variable and another.
- display what happens to one variable when another variable changes.
- collect information (Step 1).
- create new measures in the improved process map (Step 4).

Who   Managers and other employees, with the latter often identifying variables and providing the best source of data samples.

How   To construct a scatter diagram:
- Collect 50 to 100 paired samples of data relationships you wish to test and record them. For example, record the percentage of elasticity and the percentage of moisture.
- Draw the horizontal and vertical axes of a graph, making them equally long.
- Label the variable you suspect is the cause (percentage of moisture) on the horizontal axis and the variable you suspect is the effect (percentage of elasticity) on the vertical axis.
- Plot the data on the graph.
- Look for positive, negative, or nonrelationships (see Figure 19).

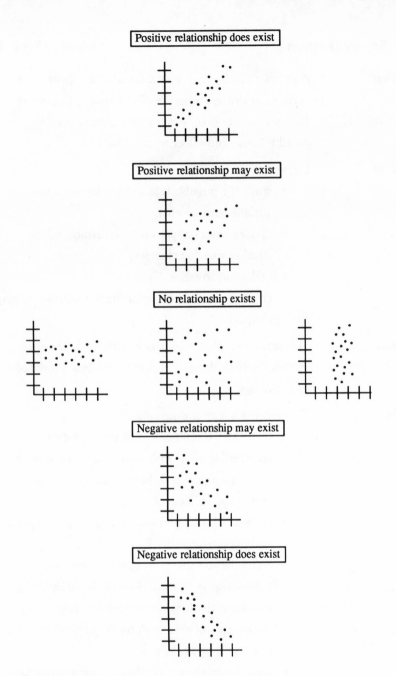

Figure 19   Scatter Diagrams

The Customer-Driven Company

Results   A scatter diagram provides:

- an analysis of your current processes, once you have converted information into measures.
- a display of the effect of preliminary improvements on a product or process. ("What's the effect on this if we change that?")
- assistance in discovering real problems when used in conjunction with cause-and-effect diagrams (see page 239).

Note: Be aware that a relationship between one variable and another does not necessarily imply a causal relationship.

For further   Kaoru Ishikawa, *Guide to Quality Control* (White
information   Plains, N.Y.: Unipub, 1988).

William Scherkenbach, *The Deming Route to Quality and Productivity* (Washington, D.C.: CEEP, 1988).

3

What   A mechanism for linking standards to the steps in a process. The matrix helps the manager to identify key employees, teams, or functional areas in a process and the standards that need to be established for them.

When   Use the standards matrix to:

- identify key steps in a process so that customer-focused standards can be developed for those steps.
- identify key teams or functional areas and the standards they must meet to be customer-focused.
- foster an environment in which standards are part of the work requirements.
- emphasize the focus on measurement and quality.
- establish standards (Step 5).

Who   Managers and other employees.

How   To use the standards matrix:

- Refer to a process map of an improved process (Step 4).
- List the key process steps across the top of the standards matrix (see Figure 20), using one or two summarizing words.
- Down the left side of the matrix, list the teams or functional areas associated with this process.
- In the appropriate spaces, write the necessary process standards for each team or area in each step of the process. These standards should meet the expectations of customers and the MARC criteria (see page 127), and should foster improvements in the quality of each step.

Results   A standards matrix provides:

- an opportunity for the manager, in collaboration with the employees, to assume responsibility for implementing the standards (from the matrix) and to review performance of the process using the standards as criteria.

Example

Process Steps

Teams/Functions	A. RECEIVE & RECORD	B. ENTER DOCUMENT	C. PHOTO COPY	D. BUDGET CONTROL
1. WORD PROCESSING CENTER	• PERFORMANCE IS UP TO STANDARD WHEN DOCUMENTS ARE ACCURATELY RECORDED W/IN 15 MIN. OF BEING RECEIVED • RECOGNIZE DELIVERER BY NAME	• DOCUMENT ENTERED ERROR-FREE, CORRECT FORMAT, WITHIN CONTRACT TIME	• PERFORMANCE IS UP TO STANDARD WHEN WPC PROVIDES 4 HOURS NOTICE ON LARGE COPYING JOBS (>500 PAGES)	• PERFORMANCE IS UP TO STANDARD WHEN ALL CROSS-CHARGES ARE SUBMITTED ON THE DAY THEY ARE INCURRED AND ARE ERROR-FREE
2. COPY CENTER			• DOCUMENT HAS ALL PAGES IN ORDER, COPYING IS DONE WITHIN 2 HOURS	
3. ACCOUNTING				• CROSS-CHARGES ARE ERROR-FREE, INCLUDING SOURCE OF CHARGE & EXPLANATION OF CHARGE, WITHIN 24 HOURS OF RECEIPT
4.				

Figure 20   Standards Matrix (Adapted from Role/Responsibility Matrix. Copyright © 1981 by The Rummler Group. Used with permission of the Rummler-Brache Group.)

What   A technique for separating clustered data points into meaningful categories to help in determining causes of problems. Stratified, categorized data can indicate significant problems, defects, trends, or relationships and present opportunities for improvement in a way that unstratified data cannot.

When   Use stratification to:
- analyze data to find opportunities for improvement.
- analyze the current process (Step 3).

Who   Managers and other employees.

How   To employ the stratification technique:
- Collect data on check sheets (see page 247) or plot scatter diagrams (see page 283).
- Determine if data can be broken down into categories. For example:
  — returns can be stratified by reason for return.
  — defects can be stratified by raw material, operator, process errors, and so on.
  — raw materials can be stratified by supplier A, B, C, and so on.
- Place the stratified data on a clean check sheet using the separate categories, or plot the data on a new scatter diagram.
- Compare the new charts with the old ones, examining both to see if there are differences when the data are broken into categories.

Results   Stratification provides:
- the opportunity to demonstrate a relationship where none was thought to exist.
- help in uncovering the probable cause of a problem.

For further   Kaoru Ishikawa, *Guide to Quality Control* (White
information   Plains, N.Y.: Unipub, 1988).

                    Statistical Quality Control Handbook (Indianapolis, Ind.:
                    AT&T, 1985).

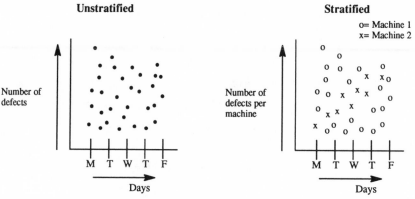

Figure 21   Machine Calibration Checks

Before stratification, no significant trends appear. After stratification, a trend can be detected on machine 2 as a subset of the larger group of data.

Before stratification, no significant trends appear. After stratification, a trend can be detected on subsets of the larger group of data.

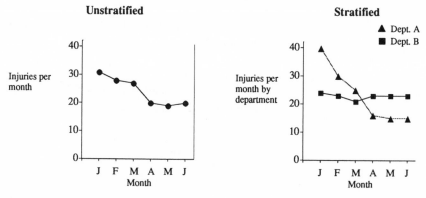

Figure 22   Major Injuries

What   A printed questionnaire requiring written responses. Questionnaires may be completed by respondents themselves or administered by a person who asks questions and writes down responses.

When   Use surveys to:

- collect information from customers (Step 1).
- learn from customers which outputs of your work unit have the greatest need for improvement (Step 1).
- identify specific customer expectations and perceptions (Step 1).
- assess and quantify conformity to customers' expectations.

Who   Managers and other employees, with the former generally designing the survey in collaboration with the latter.

How   To conduct a survey:

- Decide who and how many people will be surveyed.
- Formulate questions and test them on several colleagues or customers.
    - **Open-ended** questions yield much information and detailed responses. An example of an open-ended question is, "If you were a department manager, what would be your top priority for improvement in quality?"
    - **Closed-end** questions require only a yes or no response or simple one-word answer. A typical closed-end question might be:

Do you conduct regular staff meetings?

_____ yes _____ no (check one)

If yes, how often? _____

_____

— Closed-end questionnaires may require the respondent to make a choice. For example: How would you rate the response speed of our agents?

1	2	3	4	5

_____

Unsatisfactory	Needs improvement	Fair	Good	Excellent

Note: Open-ended questions yield high-quality data but are difficult to interpret or to use to find trends. Closed-end questions yield quantifiable data but may not thoroughly reflect respondents' feelings. It is good practice to use both, if possible.

- Decide on the method of delivery (mail, telephone, inclusion with products, services, or information).
- Encourage responses with follow-up phone calls and other reminders, because an average response rate is about 50 percent.
- Analyze results in terms of numerical averages or response categories. Look for opportunities to use such tools as check sheets (see page 247), scatter diagrams (see page 283), and histograms (see page 260).
- Draw conclusions.

Results   A survey provides:

- a determination of the needs of a large group of customers.

- trends in open-ended responses that can be helpful in identifying problems that may need attention.
- scores on items for which a numerical rating scale is provided, which can be averaged to help reach conclusions about customers as a group.
- a determination of which items in your work unit most need change or improvement or both; if you use a rating scale with the highest number reflecting excellence, the lowest scores will determine your priority.

For further information   *Process Quality Management and Improvement* (Indianapolis, Ind.: AT&T, 1989).

Seymour Sudman and Norman Bradburn, *Asking Questions: A Practical Guide to Questionnaire Design* (San Francisco: Jossey-Bass, 1983).

■ Taking the Customer's Place   Easy ●

What   Trading places with customers, both internal and external, enables you to see your own organization and efforts through the customer's eyes.

When   Use the technique of taking the customer's place to:
- learn which outputs of your work unit have the highest priority for improvement (Step 1).
- identify current gaps between customers' expectations and perceptions (Step 1).
- design improved process (Step 4).
- establish standards (Step 5).
- manage performance (Step 6).
- convert information to measures (Step 2).

Who   Managers and other employees.

How   To take the internal customer's place:
- Visit the work unit that receives output from yours, study it, observe people dealing with your outputs, and do their job using your unit's output for a day or two.

To take the external customer's place:
- Purchase the products or use the services or information of your own work unit, as your customers do.

Results   Taking the customer's place provides:
- an opportunity to experience using products, services, or information from the customer's perspective.
- a chance to turn your insights from the customer's vantage point into priorities for action in your work unit.

Acknowledgments

It is too simplistic to say a book is written. Rather, it is created, and the writing is the final rendering of the work that has been done. Literally hundreds of people have helped create this book. It is, then, not my book but our book. To these wonderful people I offer thanks and acknowledgement.

This book is about customers. It is only fitting to thank you, the reader, and the thousands of customers with whom I have worked in twenty-two years of selling, consulting, teaching, and speaking. You have let us grow and thrive. Your needs, concerns, and aspirations create the privilege of serving you.

Without The Forum Corporation, there would be no book. With deep respect I acknowledge the ability of my partners, John Humphrey, John Bray, and Don Somers, who, since we opened our doors in 1971, have kept Forum a vital, growing, and learning place—a place that strives to make a difference and, because of that, has.

Robert Chapman Wood is more than an excellent writer. He is a researcher, questioner, and stickler for details who works as hard to uncover a pertinent fact as he does to find the proper phrase to describe it. Robert and his associate Annie Meyer-Post applied themselves tirelessly to the awesome task of extracting beliefs, facts, and anecdotes from me and from Forum. Robert's organizational skills, continuing press for clarity, and his refreshing willingness to hear candid feedback created a powerful collaboration.

Peter Hillyer is my favorite writer. He also happens to be my brother, and this book represents our first opportunity to collaborate in business. Peter fine-tuned and added spark to each chapter. His mastery of the English language brought directness, economy, and fun to the book.

Reading a manuscript in its early stages sometimes means grinding through a work that is barely readable. Such was the case with *The Customer-Driven Company*. Those who invested the time to read the manuscript in its embryonic form provided me with objective and very helpful feedback that caused further refinement and, in some cases gave new direction to the book. For their persistence and insights I thank:

William Atkinson, Vice President Marketing, General Motors of Canada

Richard Atlas, Partner, Goldman Sachs and Co.

Curtis Berrien, President, The Bay Group

Richard Boyatzis, Ph.D., Professor, Weatherhead School of Management, Case Western Reserve University

Sharon Cavanaugh, CEO, Peacock Papers

Ken Dreyer, Former President, Business Interiors

Davis Dyer, Vice President and Managing Partner, The Winthrop Group, Inc.

Daniel Grieff, CEO, Bellesteel Industries

Marlene Groth, Director, Corporate Culture/Organizational Change, The Forum Corporation

Gerald Jones, Senior Vice President, The Forum Corporation

Patricia McLagan, CEO, McLagan International

Gerald Oppenheim, Vice President Sales, Peacock Papers

A. Clark Peters, CEO, Lasalle-Dietch

Jennifer Potter-Brotman, Vice President, The Forum Corporation

John Rockwell, Senior Vice President and Group Executive, retired, Booz-Allen and Hamilton, Inc.

Liz Suneby, Director of Marketing, The Forum Corporation

Hal Taylor, Director of Human Resource Development, General Motors of Canada

Gail Welch, Freelance Copywriter, Sudbury, Massachusetts

Donald Woodley, President, Compaq Canada, Inc.

I thank the many professionals of Forum who do research, partner with our customers, and continuously search for ways to translate complex concepts into practical help for these customers, for sharing both their time and insights. In particular, I would like to thank: John Humphrey for sponsoring this project and Diane Hessan for helping to shape it; John Bray, Tom Keiser, Gary Ransom, Rick Tucci, and others for always daring to break new ground; Allan Ackerman and Marlene Groth for their work in visioning; Joan Bragar and Bill Fonvielle, whose research keeps Forum fresh and continuously pushing its limits; Tom Atkinson, Lynn Bishop-Gaines, Bob Boehm, Sarene Byrne, Jane Carroll, Mary Claire Chase, Sue Cook, Paul Dredge, Allison Farquhar, Tom French, Melinda Gleason, Rick Harris, Judy Hodge, Gerald Jones, Bruce Laing, Tony Lanhston, Mike Laughlin, Bill Molloy, Mike Pickering, Jennifer Potter-Brotman, Ann Rice, Jeff Rosenthal, Carolyn Rosner, Marty Stein, Connie Steward, Amy Tananbaum, Arthur Taylor, Kelly Thomas, Denise Thompson-Smith, Carole Wong, and Bill Woolfolk for graciously responding to numerous phone and mail inquiries and providing support and insights.

The challenging and important work of managing logistics, coordinating resources, and fact-checking was performed tirelessly and most competently by Pamela Bynum and Helena Correia Olsen. While performing their regular duties, they somehow managed to "get the book out."

Other contributors to this book are: Colin Fox, CEO of Deltapoint, who in his gentle but persistent way has helped me to learn about and respect the lessons and nuances of Total Quality; Jonathan Lee, General Partner of Lee Capital and Chairman of the Board of Globe Metallurgical, shared valuable perceptions with us about that wonderful company; Dr. Thomas Malone, president of Milliken & Co., for sharing how he and others have led their company's effort to provide outstanding products and services for their customers; Art Wegner, president of the aerospace and defense group of United Technologies Corp., for special insight into the process of leading a large organization; Aleta Holub, Manager, Quality Assurance, First Chicago Corp., for sharing First Chicago's pioneering work on quality in a service institution; Paul Noakes, Vice President, External Quality, of Motorola, for crucial insights into the leadership process at that firm; Anne O'Brien for her enthusiasm and encouragement; Jennifer Potter-Brotman, Carole Wong, and Fred Jordan for their leadership in Forum's Customer-Focus Quality Capability Center; Jim Blankenship, Court Chilton, and Liz Suneby, for their marketing genius.

Selecting a publisher for this book was a new experience for me. Among a number of attractive options, Addison-Wesley stepped forward and demonstrated through their approach a strong sense of customer focus. The experience and foresight of William Patrick, our editor, has enabled us to navigate the unknown waters of our first book venture with neither anxiety nor disruption. Bill and his team of professionals have clearly sailed these waters before.

Harold and Betty Whiteley, my parents, contributed to this book in ways they will never know. The high value they placed on education helped guide me through early challenges and ultimately provided me with the confidence for this undertaking.

And finally, I thank Sharon, my wife. She is one of the most effective business people I know. Her point of view always comes from the customer's perspective, and her suggestions ring with both uniqueness and usefulness. For her insights, encouragement, and unfailing support in this and other endeavors, I consider myself most fortunate.

Index

Advanced Cardiovascular Systems, 75, 76
American Airlines, 33–36, 71, 184
American Dental Association, 92
American Express, 17
American Telephone and Telegraph
    (AT&T), 85 n.6
Amex Life Assurance Company, 187
Amica Mutual Insurance Company, 190
Apollo project, 30
Apple Computer, 151
Atari, 57
Audi, 24–25, 29
Automobile industry, 5–6, 24–25, 43. *See
    also* specific companies.

Banc One, 85 n.6, 113
Bank of America, 55, 92
Baxter Health Care Corporation, 113
BayBanks, Inc., 72–73
Beals, Vaughn, 69
Beaver, Donald L., 39–40
Bell Laboratories, 80
Bell South, 61

Benchmarking, 73, 74, 76, 80, 84, 97; and
    the Customer-Focus Toolkit, 231,
    235–36; and customer-winning
    performance, 136, 146
Bennetton, 52
Bethlehem Steel, 53
Binney & Smith, 15, 52
Boeing, 124
Boston Ballet Company, 186–87
Brainstorming, 231, 237–38, 239, 258
Bransky, Joseph R., 190
Bray, John, 109
Bread and Circus, 182
British Airways, 11, 17, 145, 173; Putting
    People First program, 32, 38 n.4,
    185; video complaint booths of, 56,
    155; vision of, 32, 38 n.4
Broad, Richard, 36
Burke, James, 24

Calsonic Corporation, 14 n.2
Campbell's Soup, 43, 85 n.6
Canon, 80

Carlzon, Jan, 43, 44
Cause-and-effect diagrams, 231, 239–42, 258
Cavanaugh, Sharon, 83–84
Cawley, Charles M., 87–112
Central Pacific Railroad, 21
Charisma, 188
Charts and graphs: cause-and-effect diagrams, 231, 239–42, 258; control charts, 232, 249–50; bar graphs, 244; line graphs, 243–44; pareto charts, 233, 268–71; run charts, 234, 278–80
Check sheets, 231, 247–48
Children's Hospital, 26
Claridges Hotel, 72
Clarry, Tony, 145
Coleco, 57
Compensation, 100–102, 110, 153, 209
Complaints, 39–40, 48, 50–51, 62–63, 66, 120; in focus groups, 52; and measurement programs, 159; receptiveness to, necessity for, 152; and stratification, 63; on videotape, 56
Conagra, 85 n.6
Congress, U.S., 139
Continental Airlines, 33
Continuous improvement loop, 80
Cookiemania, 72, 73
Coopervision Cilco, 75, 76, 85 n.4
Corey, E. Raymond, 59
Corning Works, 16, 82, 85 n.6, 101, 149, 185
Cost-benefit analysis, 172, 232, 251–52
Crandall, Robert, 33–36
Crate & Barrel, 83
Crawford-Mason, Claire, 6
Cross-functional management, 140, 142–43, 192, 198, 210
Crozier, William, 72
Customer Awareness Trips, 60–61
Customer-Focus Toolkit (Forum Corporation), 5, 18, 49, 125, 217, 219–93; and benchmarking, 231, 235–36; and brainstorming, 231, 237–38, 239, 258; and cause-and-effect diagrams, 231, 239–42, 258; and the characteristics of a customer-driven company, self-test for, 18, 219, 220–25; and check sheets, 231, 247–48; and control charts, 232, 249–50; and cost-benefit analysis, 232, 251–52; and dimensions/expectations, 232, 253–55; and focus

groups, 232, 256–57; and force field analysis, 232, 258–59; and histograms, 233, 260–62; and interviews, 233, 264–65; and nominal group technique, 233, 266–67; and pareto charts, 233, 268–71; and process evaluation, 233, 272, 274–77; and process mapping, 233, 272–77; and run charts, 234, 278–80; and sampling, 234, 281–82; and scatter diagrams, 234, 283–85; and standards matrices, 234, 286–87; and stratification, 234, 288–89; and surveys, 234, 290–92; and taking the customer's place, 234, 293; and tools for developing a vision, 37, 219, 226–28; and tools for smashing barriers to excellent performance, 219, 229–93.
Customers: final (end users), 42–44, 56, 60, 66, 124; intermediate (distributors), 42–44, 51, 66. See also Internal customers
Customer-Satisfaction Indexes (CSI), 169–71

Dean, Mary Jude, 72
Debriefing sessions, 78–79
Decision making, 27–29, 113, 173, 175, 205, 210
Delta Airlines, 54
DeltaPoint, 77
Deming, W. Edward, 6, 8
Deming Prize for Quality Control, 150
Digital Equipment Corporation, 53
Dimensions/expectations, 232, 253–55
Disney World, 173–74
Drescheck, John, 104
Ducks Unlimited, 92
Du Pre, Max, 177

Eastern Airlines, 33
Edison, Thomas A., 69
E.I. du Pont de Nemours & Company, 183
Eisner, Michael, 56
Eli Lilly, 75
Emerson Electric, 71
Emery, Allan C., Jr., 199–200
Ernst and Young, 83
Evangelista, Paul, 72–73

Federal Express, 17, 25–26, 121, 157, 158; and learning from winners, 77, 81, 85 n.6; and the Legend Makers, 72, 73
Ferguson, Homer, 37

The Customer-Driven Company

Firnstahl, Timothy, 1
First Chicago Corporation, 17, 81–82, 85 n.6
First National Bank of Chicago, 6, 165, 172, 174–75
Florida Power & Light, 85 n.6
Focus groups, 52, 56, 59, 156, 158, 165, 232, 256–57
Forbes, Malcolm, 74
Force field analysis, 232, 258–59
Ford Motor Company, 6, 29, 56, 64, 101, 211; and learning from winners, 76, 80, 81; Taurus program at, 122–23; team management at, 82
Fortune, 109, 147 n.3, 200
Forum Corporation, 3, 16, 18, 54, 103, 180–81, 186; Building Customer Focus Programs, 185; Customer Focus Executive Assessment, 9, 215–18, 220–25; research on influence, 192; survey of customer satisfaction, 49; survey of identifiable reasons for switching to a competitor, 9–10. *See also* Customer-Focus Toolkit (Forum Corporation)
Forum Europe, 109
Fuji Electric, 113, 123
Galvin, Paul, 201
Galvin, Robert, 17, 201, 203–4
General Electric (GE), 50–51, 61, 64, 100, 184; jet-engine division, 179, 180; organization of new projects at, 82
General Motors (GM), 54–55, 60, 77, 189–90; and customer-winning performance, 130, 147 n.3; and measurement programs, 173–74; set-up time at, 132
Gilbert, Robert, 95, 113
Global Metallurgical, 12, 85 n.6
Gold Award for Quality (Canada), 181
GTE, 85 n.6
Gummi Bears (television show), 56
Gunn, David, 114

Hansen, Ken, 200
Harley-Davidson, 69–71, 75, 76, 82
Harnett, Anthony, 182, 183
Harvard Business School, 7, 41, 58
Herman Miller (company), 177
Hershey, 82
Hertz, 33
Hewlett Packard, 85 n.6
Hilton Hotels, 33
Hiring, 76, 90–91, 93–98, 99, 114

Histograms, 233, 260–62
H.J. Heinz, 85 n.6
Holiday Inn, 2, 54
Holub, Aleta, 17, 175
Honda, Soichiro, 200
Honda Motors, 7, 53, 69–71, 200, 206 n.12
Honeywell, 143
Horizon, 187
Horne, Charles E., 190
Houghton, Jamie, 185
Huffy Corporation, 76
Humphrey, John, 186–87
Huntington, Collis P., 21, 22 36–37

Ignorance, 117–18, 121
Imperial Chemical Industries (ICI), 185
Inc. Magazine, 83
In Search of Excellence (Peters and Waterman), 6
Influence, 192–97, 207
Innovations, 120, 132–33, 136, 146
Inspiration, 22, 27–31, 42, 59–60, 91
Internal customers, 42–44, 52, 56, 66, 95; and customer-winning performance, 124, 130, 131, 137
International Business Machines (IBM), 11, 57, 77, 80, 82, 85 n.6, 184
Interviews, customer, 159, 165, 167–68; and the Customer-Focus Toolkit, 233, 264–65
Inventory control systems, 70
Ishikawa, Kaoru, 19 n.4, 147 n.12

Japan 12, 22, 78, 82. *See also* Management, Japanese; *specific companies*
J.D. Powers surveys, 54–55, 169
Joban Kosan, 149, 151
Joban Spa Resort Hawaiian, 149–50, 158, 161, 165, 171
Johnson & Johnson, 23–25, 28, 85 n.4
Juran, J.M., 6, 8
Just-in-time manufacturing, 4, 70, 130–31

Kaisha hoshin (Statements of basic policy), 22
Kearns, David, 85 n.1, 191
Key result areas (KRAs), 136–39
Kit control systems, 144
Kodak, 108
Komatsu, 74, 144
Kroc, Ray, 15, 140–41, 200, 201, 226
Kume, Yutaka, 188
Kyakka shoko (reflection on our shortcomings), 32

Langer, Ellen, 46
Lazarus, Charles, 39
Legend Maker program, 72–73
Leonard, Stew, Jr., 183
Levitt, Theodore, 41, 58–59
Linjeflyg, 44
Loyalty, 12, 24, 49, 59
L.L. Bean, 29, 80, 82, 136–39
Luther, David, 101

McDonald's, 15, 85 n.6, 140–41, 200–201, 226
McNeal, Danny, 16–17
Mager, Robert, 105
Malcolm Baldrige National Quality Award, 11, 26, 77, 88, 121, 139, 181, 189
Malone, Thomas, 189, 198–99
Management, Japanese, 6, 80; and American management techniques, 14, 16, 22; and "going to school" on winners, 74, 76; and *kaisha hoshin* (statements of basic policy), 22; and *kyakka shoko* (reflection on our shortcomings), 32; and suggestions from employees, 123; and "Total Quality," 19 n.4; and training programs, 104, 190; and vision, 22–23, 32
MARC criteria, 163, 286
Marcus, Stanley, 211
Marine Corps, 177
Marketing, 7, 14, 41, 135, 138; and cross-functional management, 142, 143; and measurement programs, 153, 154
Market share, 153, 164
Marous, John C., 183
Marriott, Bill, Jr., 49
Marriott Corporation, 12, 45, 49–50; 81, 82, 111–12; guest questionnaires, 45, 49, 65; Guest Satisfaction Index at, 171
Marshall, Colin, 17, 32, 145
Massachusetts Institute of Technology (M.I.T.), 104
MBNA America, 87–93
Measures, 17, 90, 92, 149–76; attribute, 125; and behavior, changes in, 174–75; customer-driven measurement, 99–100; Customer-Satisfaction Indexes (CSI), 169–71, 173; and customer-winning performance, 125,

126, 127, 128; of end results, deter-mination of, 151, 155–64; failure of, 153–55; and internal problems, 151; internal process, 156; linking of, to the organization as a whole, 171–72; quantitative questions, 160–62; reasons for, 151, 152–55; results of, communication of, 151, 155, 172–76; and standards, establishment of, 162–63, 176; and "surrogates" for quality, 163–64; variable, 125. *See also* Questionnaires; Surveys.
Merck Pharmaceuticals, 190–91
Meridian Bancorp, 56
Methodist Hospital, 12–13
Metropolitan Life, 102
Milliken, Roger, 199
Milliken & Company, 77, 85 n.6, 189, 190, 191, 198
Mindfulness (Langer), 46
Minnesota Mining & Manufacturing (3M), 82, 85 n.6, 151
Motorola, 11, 16, 17, 191, 201–4; and learning from winners, 77, 81, 85 n.6
Mystery shoppers, 156, 165, 168–69

National Broadcasting Company (NBC), 6
National Quality Forum, 89
Naumann, William, 82
Navy, U.S., 177–78, 179
Neiman Marcus, 211
Nestlé USA, 85 n.4, 85 n.6
New Pig Corporation, 39–40
Newport News Shipbuilding and Dry Dock Company, 21, 22, 23, 36–37
New York Times, The, 54
New York Transit Authority, 114
Nintendo Company, 57
Nippon Telephone, 32
Nissan, 14 n.2, 50, 74, 119, 188
Noakes, Paul, 81, 201–4
Nolan, Sarah, 187
Nominal group technique, 233, 266–67
Non-profit organizations, 82, 92
Northwest Airlines, 33, 36
Nyland, Larry, 81

Ohno, Taiichi, 129–31
Olsen, Ken, 53

Pan American Airlines, 33

Paperwork Reduction Act, 139
Partnerships, with customers, 61–64
Peacock Papers, 83–84
Peck, M. Scott, 80
People Express Airlines, 33
Peterson, Donald, 211
Piedmont Airlines, 33
Pratt & Whitney, 178–82, 184
Problem-solving, 4, 75, 145, 181, 210; and customer-winning performance, 120, 145, 146; and "going to school" on winners, 77; and the identification of problems, 76, 79; and Quality Circles, 151; selection of methods for, 125; self-test regarding, 222; six-step procedure for, 120
Process: analysis, selection of methods for, 125; design of an improved, 125–26; evaluation, 233, 272, 274–77; mapping, 125, 233, 272–77; standards, 126–27
Procter & Gamble, 85 n.6, 184
Productivity norms, 99–100
Promotion, 76, 100, 102–3, 179

Questionnaires, 66, 159, 165–67, 192; Marriott, 45, 49, 65. See also Surveys

Radio Corporation of America (RCA), 6
Ramich, Joel, 149, 172, 185
RATER structure, 47, 159
Recession, 34, 43, 44
Red Lobster, 45
Reliability, 47, 54, 118, 159–60, 253
Republic Airlines, 33
Research and Development (R&D), 142, 143
Rewards, 100–103, 114, 217
Riley, Harry, 72–73
Road Less Traveled, The (Peck), 80
Robinson, Susan, 124
Rohm & Haas Bayport, 95, 112–13
Rowe, Darrell, 60
Rubbermaid, 118
Runyon, Damon, 13–14
Russell, Larry, 172
Ryder Systems, 61

Sales and Marketing Magazine, 138
Sales trends, electronics monitoring of, 52, 66

Sampling, 234, 281–82
Satisfaction Guaranteed Eateries, 1
Scandinavian Airlines (SAS), 43–44
Scatter diagrams, 234, 283–85
Schonberger, Richard, 117
Scoggin, Daniel R., 141
Sea Land Corporation, 170
ServiceMaster Company, 109, 199–200
Sewell, John, 107
Sheth, Jagdish N., 41
Shinto, Hisashi, 32
Siefkin, William, 183
Sierk, James, 76
Singapore Sheraton Towers, 1, 3
60 Minutes, 24
Sloan, Alfred, 189
Smith, Fred, 17, 25
Smith & Hawken, 72, 73
Southern Pacific Railroad, 21
Standards, process, 126–27
Standards matrices, 234, 286–87
Steuban Glass, 16, 17
Stew Leonard's, 80–81, 183
Strategic Planning Institute, 7–8
Strategy, 7–8, 27, 34–36
Stratification, 234, 288–89
Subaru of America, 52
Sugiura, Hideo, 7
Sundry, Arthur, 201, 203
Surveys, 40, 66, 125; and the Customer-Focus Toolkit, 234, 290–92; J.D. Powers, 54–55, 169; mail, and measurement programs, 166-66, 168; and post-purchase assessments, 56; and potential products, 59; and the RATER structure, 47; by Technical Assistance Research Products (TARP), 40; and the voice of the customer, 46, 47, 49, 56. See also Forum Corporation

Taylor, N. Powell, 51
Teams, 128, 129–30, 143, 181, 187, 191–97; and the influence test, 193–97
Technical Assistance Research Products (TARP), 10, 40, 55
Technology, 52, 69, 119, 126, 132, 133, 142
Tennant, Inc., 85 n.6
Tenneco, 23, 36

Texas A&M University, 47
Texas Air, 33
Texas Instruments, 85 n.6
TGI Friday's, 141
The Limited, 83, 108
Toni, Robert, 75, 76, 77, 85 n.4
Toolkit. *See* Customer-Focus Toolkit (Forum
    Corporation)
Total Product Concept, 58–59
Total Quality Control, 138
Toyoda, Kiichiro, 119
Toyota, 74, 118, 119, 128, 130–32, 147 n.3
Toys 'R' Us, 39, 63
Training, 61, 104–7, 190–91, 202; and
    customer champions, 91, 104–7, 110,
    114; and customer-keeping vision,
    104, 105–8, 190; and customer-
    winning performance, 137
Travelers Insurance, 40
Trust, 47, 64, 193, 196
Turner, Fred, 140-41
Tylenol incident, 23–25, 29

United Airlines, 33, 36
United Technologies Corporation, 133–36,
    178, 180
USAir, 33

Value, 48, 64, 126, 181, 205
Vanderzwan, Michael C., 190–91
Videotapes, 52, 55, 56, 61, 79; complaint
    booths, and British Airways, 56, 155;
    and measurement programs, 150, 161
Visible management, 140, 140–44, 146
Vision, customer-keeping, 15, 28, 199–205,
    210; actions points for, 37; and the
    anatomy of a living vision, 26–29;
    and the barrier-free company, 140;
    and customer champions, 95, 100,

114; and decision making, 27, 28–29;
    development of, 18, 21–38, 186; and
    measurement programs, 172;
    promotion of, 181, 184–88; self-test
    regarding, 221; and "staying the
    course," 198; and training programs,
    104, 105–8, 190
"Visual management," 78
Volkswagen, 5
Vorhauser, 108

Walesa, Lech, 177
Wallace Company, 121
Wall Street Journal, 152
Wal-Mart, 7
Walt Disney Company, 7, 56, 106–7, 108,
    118. *See also* Disney World
Washington Post, The, 179
Watson, Thomas J., Sr., 184
Wegner, Arthur, 177–78, 179, 180, 184
Welch, Jack, 184
Wells, Frank, 107
Westinghouse, 16, 71, 81, 183
Weyerhaueser, 85 n.6
Whitman Corporation, 82
World War II, 3, 16, 22, 23, 36, 74, 130
Worthington Industries, 101
Wright, Orville, 33
Wright, Wilbur, 33

Xerox, 64, 77, 124, 132–33, 138; and
    learning from winners, 71, 76, 80,
    82, 85 n.6; training programs at, 191

Yamada, Susumu, 150
Yoshida, Sidney, 117, 147 n.2

Ziglar, Zig, 93

Reader's Questionnaire

You are my customer. Would you please take some time to let me know whether this book has met or exceeded your expectations? It should take no more than five minutes to fill in and return this questionnaire.

Please answer the questions below. (Photocopy the pages if you wish.) Select the number on a scale of 1 to 7 that best represents your response to each question.

1. The use of words in this book is at a level that is appropriate for readers with my background and education. (Circle one)

1	2	3	4	5	6	7
strongly disagree			neither agree nor disagree			strongly agree

2. I believe I can rely on the information in this book.

1	2	3	4	5	6	7
strongly disagree			neither agree nor disagree			strongly agree

3. The tables and illustrations used in this book supplement the text well.

1	2	3	4	5	6	7
strongly disagree			neither agree nor disagree			strongly agree

4. This book is laid out in a way that makes it easy for me to find the information I need.

1	2	3	4	5	6	7
strongly disagree			neither agree nor disagree			strongly agree

5. The tone of this book is just about right: not too academic and not too casual.

1	2	3	4	5	6	7
strongly disagree			neither agree nor disagree			strongly agree

6. The anecdotes and examples illustrate well the points that are being made.

1	2	3	4	5	6	7
strongly disagree			neither agree nor disagree			strongly agree

7. This book includes much information that is new to me.

1	2	3	4	5	6	7
strongly disagree			neither agree nor disagree			strongly agree

8. I find I can apply ideas from this book to my work.

1	2	3	4	5	6	7
strongly disagree			neither agree nor disagree			strongly agree

9. I expect that I will continue to use this book as a reference work or source of ideas.

1	2	3	4	5	6	7
strongly disagree			neither agree nor disagree			strongly agree

10. I have recommended this book to others, or probably will.

1	2	3	4	5	6	7
strongly disagree			neither agree nor disagree			strongly agree

11. Please indicate the extent to which this book has met your expectations overall.

1	2	3	4	5	6	7
strongly disagree			neither agree nor disagree			strongly agree

12. In this space, please describe how you will apply the information in this book to your work situation.

13. In this space, please tell me at least one thing that would make this book more useful to you.

Please send your completed questionnaire to:

Richard C. Whiteley
The Forum Corporation
One Exchange Place
Boston, MA 02109